Series/Number 07-110

CALCULUS

GUDMUND R. IVERSEN
Swarthmore College

SAGE PUBLICATIONS
International Educational and Professional Publisher
Thousand Oaks London New Delhi

For information address:

SAGE Publications, Inc.
2455 Teller Road
Thousand Oaks, California 91320
E-mail: order@sagepub.com

SAGE Publications Ltd.
6 Bonhill Street
London EC2A 4PU
United Kingdom

SAGE Publications India Pvt. Ltd.
M-32 Market
Greater Kailash I
New Delhi 110 048 India

Printed in the United States of America

Library of Congress Catalog Card No. 89-043409

Iversen, Gudmund R.
Calculus/Gudmund R. Iversen.
p. cm.—(Sage university papers series. Quantitative applications in the social sciences; no. 07-110)
Includes bibliographical references.
ISBN 0-8039-7110-9
1. Calculus. I. Title. II. Series.
QA303.I94 1996
515—dc20 95-41774

00 01 10 9 8 7 6 5 4 3 2

Sage Production Editor: Gillian Dickens
Sage Typesetter: Andrea D. Swanson

When citing a university paper, please use the proper form. Remember to cite the current Sage University Paper series title and include the paper number. One of the following formats can be adapted (depending on the style manual used):

(1) IVERSEN, GUDMUND R. (1996) Calculus. Sage University Paper series on Quantitative Applications in the Social Sciences, 07-110. Thousand Oaks, CA: Sage.

OR

(2) Iversen, G. R. (1996) *Calculus* (Sage University Paper series on Quantitative Applications in the Social Sciences, series no. 07-110). Thousand Oaks, CA: Sage.

CONTENTS

Series Editor's Introduction v

1. Introduction 1

Mathematical Functions 2
Graphs of Functions 3
Continuity 5
Calculus 6
Differentiation and Integration 7
Duality 8

2. Differentiation 9

Introduction 9
Definition of the Derivative 9
Simple Examples 14
The Derivative of a Sum of Two Functions 15
Other Functions 16
Derivatives of a Product and a Quotient of Two Functions 21
The Chain Rule 22
Different Notations 23
Higher Derivatives 25
List of Derivatives 28

3. Integration 29

Introduction 29
The Definite Integral 30
Area and Slope 34
Adding Small Rectangles 37
The Fundamental Theorem of Calculus 39
The Integral as a Function of a Limit 42
Specific Integrals 44
Integration Methods 46
List of Integrals 48

4. Applications 49
Maximum and Minimum 49
Integration 55
Distance, Speed, and Acceleration 60
Applications in Statistics 66
Applications in the Social Sciences 77
Conclusion 80

References 81

About the Author 82

SERIES EDITOR'S INTRODUCTION

In the history of science, great discoveries sometimes occur in mysterious pairs. Two independent researchers may suddenly solve a major problem that has been intractable for years, even centuries. One leading example is the discovery of the radio. Another is that of calculus, simultaneously developed by Newton and Leibniz in the 17th century.

A central field within mathematics, calculus can be thought of as a set of rules, or operations, performed on mathematical functions. The two basic operations are differentiation and integration. Differentiation tells us the change in a function when a variable, x, changes. To differentiate means to take a derivative, and a derivate says how much change there will be in a function, say y, as x changes by a limited amount. The integration of a function leads to another function: Integration, a sort of reverse differentiation, helps us find the area under a curve.

With a lucid expository style, Professor Iversen unfolds the methods of differentiation in Chapter 2. After a definition and some examples, he explains the derivatives for a sum of two functions, other functions, and a product and a quotient of two functions. He then goes on to the chain rule, different notation, higher derivatives, and a list of derivatives. Integration is the focus of Chapter 3, where he includes discussion of definite and indefinite integrals, the fundamental theorem of calculus, and integration as a function of limit. He continues with specific integrals, such as integration of a polynomial function and the sine function. With regard to integration methods, he notes that two are commonly used—integration by substitution and integration by parts—and offers an example of the former. The chapter concludes with a list of integrals.

Most brief calculus introductions fall short when it comes to providing the reader with applications. Iversen's book easily escapes this criticism. In the final chapter, there are numerous worked examples of differentiation, such as the largest area a fence can enclose, the shortest time from a lifeguard to a swimmer, and the largest can of soup. Examples of integration include money growth through compound interest, the volume of a sphere, and a distance function. As well, there are direct applications from statistics concerning, for example, the expected value, the population mean, and p values. With respect to the last, we know that the area under a curve can stand for the probability of a t statistic landing in a given interval on the x-axis, and that this can be represented through calculus. Thus, calculus underlies the inferential math of hypothesis testing. Further, Professor Iversen shows us that the least squares solution of regression analysis rests on calculus. In particular, a slope coefficient is a derivative (or partial derivative in the multiple regression case) indicating the instantaneous change in the dependent variable for a given change in the independent variable.

Beyond the statistical applications, examples are also given in which certain social or political processes are modeled directly from calculus, such as arms races and the famous Richardson equations. All practicing social scientists need to have an appreciation of calculus, the foundation of so much that we do. The paper at hand, written gracefully and simply, provides that appreciation. After reading it, data analysts will understand the meaning behind hitherto rote calculations of the computer. In addition, they may go on to explore calculus models of human behavior.

—*Michael S. Lewis-Beck*
Series Editor

CALCULUS

GUDMUND R. IVERSEN
Swarthmore College

1. INTRODUCTION

Calculus has been one of the major branches of mathematics for more than 300 years. The foundations of calculus trace back to work done during the second half of the 17th century by Isaac Newton and Gottfried Wilhelm von Leibniz, independently of each other. In turn, they built on work done by earlier mathematicians. We can see one example of the importance of calculus today in the fact that mathematicians have decided to make it the culmination of mathematics for the better high school students and the introduction to mathematics for large numbers of college students.

Calculus is based on sophisticated mathematical ideas of limits and infinitesimally small quantities. We will not be able to do full justice to these ideas in this short introduction to calculus, but even the intuitive explanations of these ideas given here will aid in the understanding of calculus. For more rigorous introductions to calculus, the interested reader should consult textbooks on the high school or college level.

One simplified way to look at calculus is to say that it consists of operations that can be applied to mathematical functions. For example, $f(x) = x^2$ is a mathematical function of the variable x that assigns the square of the variable to any value of the variable. We can apply calculus methods to this function. A more mathematical view of calculus is that it is the study of functions via linearization. A more elaborate and perhaps more understandable view of calculus is provided in the *McGraw-Hill Encyclopedia of Science and Technology* (Parker, 1992):

> The branch of mathematics dealing with two fundamental operations, differentiation and integration, which are carried out on functions. The subject, as traditionally developed in college textbooks, is partly an elementary development of the purely

---

AUTHOR'S NOTE: I am very grateful to David Rosen and two referees who tried to keep me mathematically honest and who provided many valuable comments on an earlier draft. I am also very grateful to Bobbie Iversen for reading an earlier draft from the neophyte's point of view. Any remaining ambiguities and errors are all my own.

theoretical aspects of these operations and their interrelations, partly a development of rules and formulas for applying calculus to the standard functions . . . , and partly a collection of applications. (Vol. 3, p. 146)

The reason calculus is important beyond mathematics itself is for the understandings it provides of phenomena in the world outside mathematics that are represented by mathematical functions. Calculus has other uses as well. For example, Chapter 4 shows how calculus has made it possible for statisticians to develop methods such as regression and analysis of variance.

Mathematical Functions

It is necessary that we have a good sense of what the term *mathematical function* stands for before we discuss what calculus is and what it can be used for. Suppose we have a square room. How much carpeting do we need if we want to cover the floor of such a room? The answer is that if the sides of the room are 8 feet, then we need 8 times 8, or 64, square feet of carpet. If the sides of the room are 12 feet, then we need 12 times 12, or 144, square feet of carpet. Whatever the sides of the room are, we can always figure out the area of the floor by this kind of multiplication.

We can also create a general expression that enables us to find the area for any dimensions of the room. Instead of using a specific number of feet, we can call a side of the room x feet. The area of the floor then becomes x feet times x feet, or x^2 square feet. Thus, the area of the floor *depends on* the length x of the side; we can say that the area is a *function* of the side. One way to write the area mathematically is to write it as a formula:

$$f(x) = x^2 \tag{1.1}$$

In this notation, the letter f stands for function, and the parentheses indicate a function of the variable named inside the parentheses, in this case x. The function itself is the square of x. We do not have to use the letter x for the variable. We could have called the variable by some other letter, for example t, and we could then write the same function as $f(t) = t^2$. Now that we have this function, we can substitute any value of the variable for a specific room to find the area of the floor of that room. When the side of the room is $x = 8$ feet, then the area becomes $f(8) = 8^2 = 64$. Similarly, $f(12) =$ 144, and so on.

Thus, a mathematical function $f(x)$ is some mathematical expression involving a variable x. When we choose a value of x and substitute this value into the function, we get a value of the function itself. It is like putting a number into a box and turning a crank to produce another number. For example, when we put the number 8 into the box labeled $f(x) = x^2$, out comes the number 64.

Mathematical functions take many forms. The function we have looked at is an example of a *power* function. This is because it involves raising the variable to some exponent, or power. When we have the sum of several terms with the variable raised to different exponents we get a *polynomial* function. As an example of a polynomial function we have

$$f(x) = 3x^4 - 12x^3 + 7x^2 + 2x - 5$$

Without going into their definitions, other functions in the mathematical repertoire are *trigonometric* functions such as $f(x) = \sin x$ and $f(x) = \cos x$. The basic exponential function is $f(x) = e^x$, and the basic logarithmic function is $f(x) = \ln x$. These other functions look more complicated, perhaps, but they do the same type of thing as the function for the area of a square room. We provide a value of the variable x, and for this value of x we get a corresponding value of the function of x. For example, for x equal to 3 for the logarithmic function we get $f(3) = \ln 3 = 1.0986$, to four decimals.

Graphs of Functions

Any time we have a mathematical function, we can display the function in a graph. That means there are two ways to represent a function: Either we can use the formula or we can use the graph. Sometimes it is better to use the formula, and sometimes it is better to use the graph when we work with a mathematical function.

We find the graph by choosing many different values of the variable, then for each of those values of the variable computing the corresponding value of the function. Let us go back to Equation 1.1 and find a graph of that function. Table 1.1 shows a set of values of the variable ranging from $x = 0$ to $x = 2.5$. For different values of the variable x the table shows the corresponding value of the function $f(x)$.

To find the graph of the function, we first have to draw a coordinate system with one horizontal x-axis and one vertical y-axis. Along the horizontal axis we mark off the values of the variable x, and along the vertical axis we mark off the values of the function $f(x)$. For any pair of

TABLE 1.1
Chosen Values of x and Computed Values of $f(x) = x^2$

x	0	0.2	0.4	0.6	0.8	1.0	1.5	2.0	2.5
$f(x)$	0	0.04	0.16	0.36	0.64	1.00	2.25	4.00	6.25

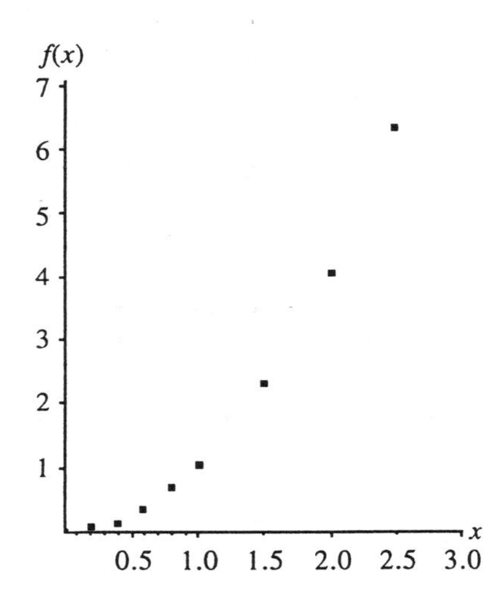

Figure 1.1. Points on the Graph of $f(x) = x^2$

numbers in a column in the table we can then plot in the graph the point that corresponds to the two numbers. Figure 1.1 shows the points displayed in Table 1.1.

Next, we choose more values of x between the values we have already chosen and compute the corresponding values of $f(x)$. The points we get this way will eventually fill in all the spaces between the points already shown in Figure 1.1. The end result is a smooth curve through all the points. Figure 1.2 shows this curve for values of x from 0 to 2.5. The figure also shows a few of the original points, marked as squares.

The reason graphs of functions are so useful in calculus is that several aspects of calculus have strong geometric interpretations. We can often

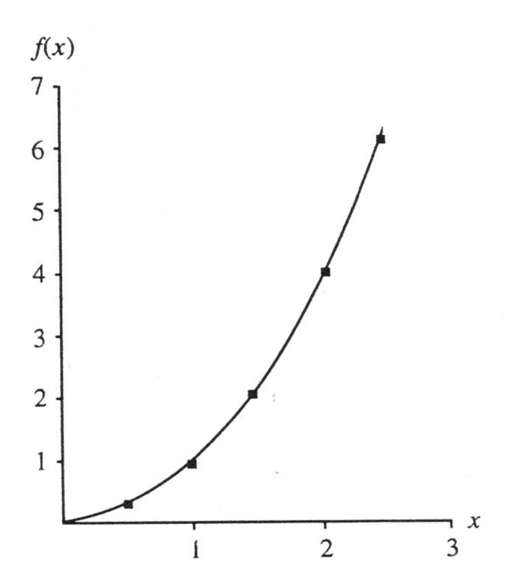

Figure 1.2. Graph of the Function $f(x) = x^2$ Together With a Few Points

make use of these geometric interpretations of calculus to learn more about the real-world phenomena that led to the graphs in the first place.

Continuity

What makes it possible to apply calculus to the function in Figure 1.2 is that the function $f(x)$ is *continuous;* that is, $f(x)$ has a value for any value of x within a given range. Mathematicians have a very specific requirement for a function to be continuous. They say that a function is continuous within a range of values if it can be evaluated at any value on the real number line within that range of values. Loosely speaking, it may be enough for the moment to say that we think of a function as continuous if we can evaluate it at any value of the variable that lies between any two chosen values of the variable. For a function to be continuous, the variable(s) in it must also be continuous, or be able to take on any value within a range. Simple but familiar examples of continuous variables are time, increase in height or weight, and speed. When we have phenomena in the real world that can be described by continuous variables and continuous functions of these variables, we can use calculus operations on those functions.

Sometimes we have variables that are not continuous, and we say that these variables are *discrete*. Loosely speaking, we have a discrete variable when the variable is such that between any two adjacent values we cannot find another value. The number of children in a family is a discrete variable because we can have a family with two children and another with three children, but obviously there are no families for which the number of children lies somewhere between two and three. For discrete variables it is harder to use calculus. More often, for discrete variables we use mathematical methods known under the label of discrete mathematics.

It is not enough to require a variable to be continuous. To use calculus methods we also need the function $f(x)$ of the variable x to be well behaved, without sharp bends or jumps. For example, if the variable x is continuous, then we can apply calculus to the function $f(x) = x^2$. All the functions discussed in this volume are well behaved, except for the function displayed in Figure 4.6. That figure shows a graph of the function $f(x) = |x|$. This is the absolute value function. For example, if $x = 3$ then $f(3) = 3$, and if $x = -3$, then $f(-3) = 3$ as well. From the figure we see that at $x = 0$ the curve of the function makes a 90 degree sharp bend. This means we cannot apply the calculus method known as differentiation of the function at that particular value of x. For all other values of x we can differentiate the function.

Calculus

The appeal and strength of calculus are that when we apply calculus to phenomena in the real world that are expressed in mathematical equations, often the result is that we learn new things about these phenomena that otherwise were not evident. Physics and engineering may be the two fields that have benefited the most from such applications of calculus. The social sciences, with the possible exception of economics, have been much less successful in their use of calculus.

For example, by using calculus on a certain function we can get the rate of change of a variable. This means we can study the rate of change of anything from the position of a car on a freeway to the position of a spacecraft in outer space to the size of a population. Similarly, if we know the rate of change, then we can use calculus and get back to the original function. Furthermore, if we can study change, then we can also find at what value of a variable the function has no change. That can correspond to a maximum or a minimum value of the function.

To take two examples from statistics, discussed further in Chapter 4, one part of calculus forms the basis for the entire least squares theory involving both regression and analysis of variance. We can view another part of calculus as an addition of many small elements between two boundaries to get the area under a curve. This corresponds to how we find a p value in statistical hypothesis testing.

Learning calculus includes learning the methods of calculus and applying them to various mathematical functions. Sometimes this can be very tedious and difficult, and it may take much experience and practice to arrive at correct answers. This monograph is not a textbook in calculus, and we consider only some of the simpler methods of calculus. The aim here is to provide an understanding of the basic ideas of calculus and to give examples of how they work.

Differentiation and Integration

Calculus consists of two separate parts. As mentioned in the earlier quotation, we call one set of operations *differentiation* and the other *integration*.

Differentiation

The basic idea of differentiation is that it helps us to study how fast a function changes when we change the x values. For example, consider the function $f(x) = x^2$ again. If x changes from 1 to 2, then the function changes from $1^2 = 1$ to $2^2 = 4$, a difference of 3. Similarly, if the variable changes by the same increase from 3 to 4, then the function changes from 9 to 16, a difference of 7. We see that the function changes faster—for the same change in the value of x—when it is evaluated at larger (positive) values of x. To study the change in the function more systematically, we can differentiate the function instead of considering specific examples.

When we differentiate the function $f(x)$, we get the function $f'(x) = 2x$. We use this notation with the prime symbol to remind us that we started with a function called $f(x)$. In the next chapter, we see how $f'(x)$ tells us about how fast the function changes.

Integration

The basic idea of integration is that when we integrate a function, we get another function. In the simplest case, this new function can be used to find the area under a curve. Sometimes this new function can be used to find volumes of solid objects, and other times integration has more sophisticated applications.

This means that to apply calculus we have to learn two sets of operations. Differentiation is discussed in Chapter 2 and integration in Chapter 3. What determines whether we want to differentiate a function or to integrate a function depends on the problem at hand. Usually it is obvious from the problem whether we differentiate or integrate.

Most people learning calculus find differentiation to be easier than integration. The reason is that when we differentiate a function, it is as if we have a question and we want to find the answer to this question. When we integrate a function, however, it is as if we have the answer and we want to find what the question is that could have produced that answer. For one thing, a given question usually has one correct answer, but an answer can result from several different questions.

Duality

After going through the basic ideas of differentiation and integration, we can see that there is an important duality between the two. Suppose we know that the function $f'(x) = 2x$ is the function we find after differentiating another function. Is there any way we now can find that original function? Yes, there is, and this is where integration comes in. When we integrate the function $f'(x) = 2x$, we get back the original function $f(x) = x^2$ (with a minor modification that need not concern us at this time). Thus, calculus makes it possible for us to go back and forth between the two functions.

Figure 1.3 illustrates this inverse relationship between the two operations. The figure shows that if we start with the function $f(x)$ and differentiate that function, then we get a new function $f'(x)$. If we integrate the function $f'(x)$, then we can get $f(x)$ back again.

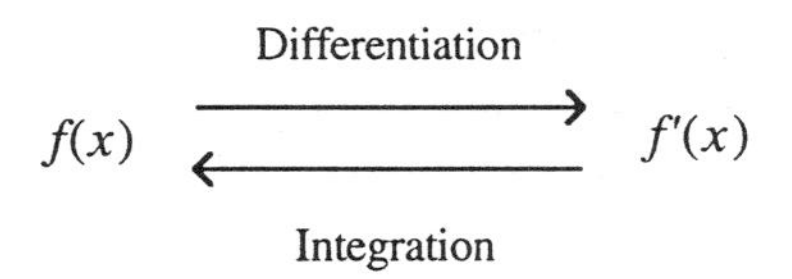

Figure 1.3. Differentiation and Integration as Inverse Operations

This inverse relationship between differentiation and integration may appear surprising, even after we have learned some calculus. We see in Chapter 2 that differentiation of a function amounts to finding how steep the tangent to the curve of the function is for each value of x. We see in Chapter 3 that integration amounts to finding the area under the curve of a function between any two values of the x variable. It is not immediately obvious why there exists such an inverse relationship between the slope of the tangent to a curve and the area under that curve, but it may help to keep this duality in mind as we explore the two parts of calculus.

2. DIFFERENTIATION

Introduction

Differentiation of functions makes up one part of calculus. When we differentiate a mathematical function, we get another mathematical function. For example, if we have the function $f(x) = 3x^2 + x - 5$, then it turns out that differentiating this function gives us the function $f'(x) = 6x + 1$. If, as another example, $f(x) = \sin x$, then $f'(x) = \cos x$. Sometimes we represent the function $f(x)$ by the letter y, and in that case the function we get after differentiating is represented by y'. To differentiate a function is the same as finding the derivative of the function. From the definition of differentiation and from the examples given below, we see why it may be useful to differentiate a function and what we can learn about a function and thereby also about the phenomenon the function represents.

Definition of the Derivative

Look at the graph of the function $f(x)$ in Figure 2.1. The graph shows a smooth curve. Because there are no "jumps" in the curve, we see that the function itself is continuous. When we have such a continuous function, we can differentiate the function.

We have picked two points a and b on the horizontal x-axis. When we substitute these two values of x into the function, we get the two values $f(a)$ and $f(b)$ of the function itself. These two values are marked on the vertical y-axis. The two corresponding points on the curve itself are denoted A and B. Finally, there is a straight line drawn through those two points on the curve.

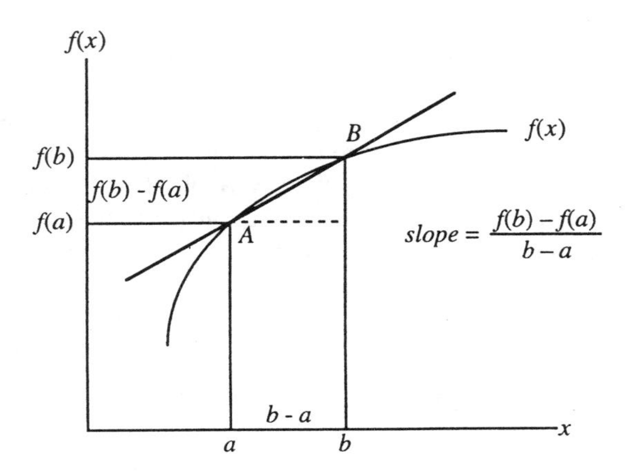

Figure 2.1. A Line Through Points A and B, With Slope

The question we now ask in calculus is this: What happens to this line through A and B as the value b moves closer to the value a? As the value b moves closer to the value a, then on the y-axis $f(b)$ will move closer to $f(a)$, and on the curve the point B will move closer to the point A. As this happens, the line through A and B comes closer to becoming the tangent line to the curve at the point A.

The figure also has an expression for the slope of the line. The slope tells us how steep the line is, and it is popularly defined as the "rise" divided by the "run." Because the line becomes close to being the tangent line, the slope of the line becomes close to being the slope of the tangent line to the curve. The derivative of the function $f(x)$ at the point A is now equal to the slope of the tangent line at that point.

To show how we formally define the derivative, we first change the notation, as in Figure 2.2. Let the value a be denoted by the generic value x and the value b by $x + \Delta x$. That means we can write the distance between the two points on the horizontal axis as Δx. At the same time, we can write the values of the function at the two points as $f(x)$ and $f(x + \Delta x)$. This means that the distance between the two corresponding points on the vertical axis becomes $f(x + \Delta x) - f(x) = \Delta y$. The slope of the line through A and B in the figure becomes, in this new notation,

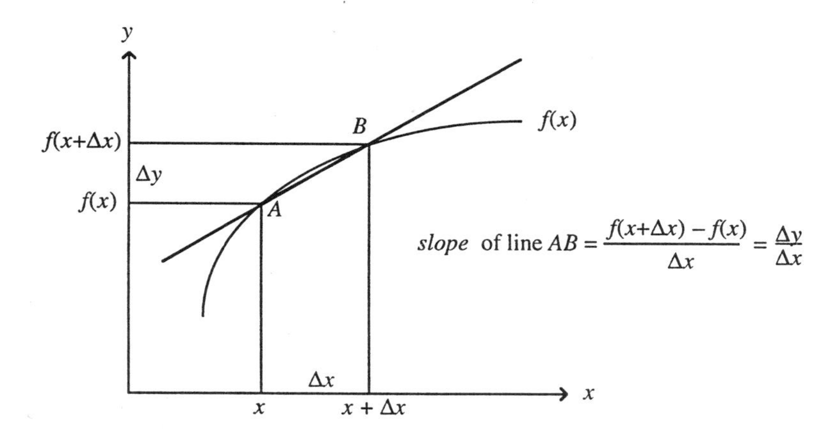

Figure 2.2. The Derivative as the Limit of a Slope

$$\text{slope} = \frac{f(x + \Delta x) - f(x)}{\Delta x} = \frac{\Delta y}{\Delta x} \qquad (2.1)$$

Now we want to see what happens to the slope of this line through the points A and B as the distance Δx on the horizontal axis gets smaller. This means we want to see what happens to this fraction when the denominator Δx gets smaller. In the limit, the distance Δx becomes zero. At the same time, the difference in the numerator also gets smaller, and in the limit it becomes zero as well. In the limit, therefore, both the numerator and denominator become equal to zero. The fraction 0/0 is undefined and therefore not of much use; fortunately, what we want to know is what happens to the fraction *just before* both numerator and denominator become zero.

This is a very sophisticated issue in mathematics, and we cannot hope to do justice to the issue here. For the moment, it suffices to say that we now define the derivative of the function $f(x)$ at the point x as the limit of Equation 2.1 as the distance Δx goes to zero. Thus, we define the derivative $f'(x)$ of the function $f(x)$ as the limit of a fraction as the denominator gets closer to zero. That is,

$$f'(x) = \lim \frac{f(x + \Delta x) - f(x)}{\Delta x} \qquad \text{as } \Delta x \to 0 \qquad (2.2)$$

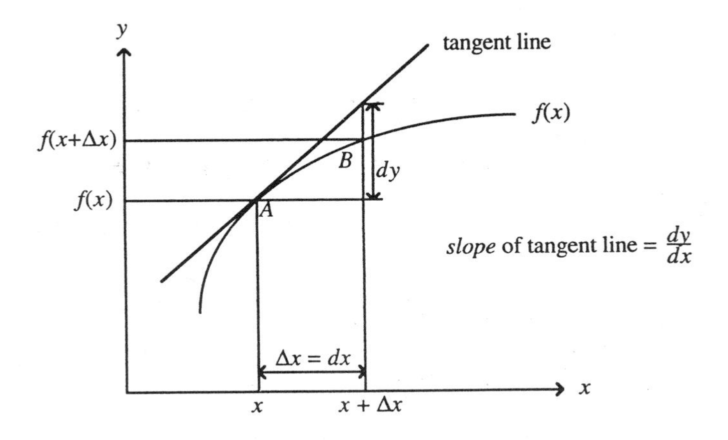

Figure 2.3. The Derivative as the Slope of the Tangent Line

The right-hand side says that we are looking at the limit of this fraction as the quantity Δx goes to 0.

As Δx gets smaller, the point B gets closer and closer to the point A. This means that the line through the two points A and B gets closer and closer to the line that is the tangent to the curve at A. Thus, by the definition of the derivative, the derivative of a function at a value of x equals the slope of the tangent line to the function at that value of x. Later we will see that the derivative can have other meanings. For example, in Chapter 4 we see that the derivative of a certain function gives us the velocity of a moving object.

Notation

Figure 2.3 shows the tangent line to the curve at the point A. This line has a certain slope. In the spirit of writing the slope of a line as a fraction, where the numerator is the "rise" and the denominator is the "run," we now write $f'(x)$ as such a fraction. Any number can be written as many different fractions. For example, 0.25 can be written as the fractions $1/4$, $2/8$, or $4/16$, among others. As soon as we choose either numerator or denominator, the other quantity becomes fixed. Let us now choose the denominator as a small Δx, which we will call dx. Figure 2.3 shows that for a given denominator dx, the corresponding numerator for the slope becomes the quantity dy.

With this new notation we now can write the derivative $f'(x)$ as

$$f'(x) = \frac{dy}{dx} \tag{2.3}$$

From this it follows that we can also write the numerator dy as

$$dy = f'(x)dx \tag{2.4}$$

With this notation we have introduced the *differential* of y, or of $f(x)$, as the quantity dy. This is the notation introduced by Leibniz in the year 1684. It turns out that this notation using dy and dx is very useful for certain purposes.

An Example. As an example of a derivative and what we can learn about the function from the derivative, let us look at the function $f(x) = 3x^2 + x - 5$. We find that the derivative of this function is another function $f'(x) = 6x + 1$. For any value of x we can now find the slope of the tangent to the curve displayed by $f(x)$ by substituting for x in the derivative $6x + 1$. For example, if $x = 0$, then $f'(0) = 1$. This means that the slope of the tangent to the curve at that value of x is 1; that is, the tangent line here is a 45 degree line. If $x = 10$, the slope of the tangent line is $6(10) + 1 = 61$. Thus, at that point, the tangent is a very steep line.

If x is large and positive, then we see that $6x + 1$ is large and positive, and so the slope of the tangent is large and positive. When the slope of the tangent is large, then the tangent is a steep line. This means that the curve rises quickly. On the other hand, if $x = -10$, then the slope of the tangent line is -59. If x is large and negative, then $6x + 1$ is large and negative, so the slope of the tangent is large and negative, meaning that the curve comes down quickly.

Finally, the derivative $6x + 1$ equals 0 if $x = -1/6$. Thus, at that point the slope of the tangent line is 0, meaning that the tangent line is horizontal. This means that the curve flattens out at that point. All of this is not very surprising, because the original function is the equation of a parabola, and the curve is ∪-shaped with the bottom at $x = -1/6$. From all of this we see that the derivative of a function can be helpful when we want an idea of what the graph of the original function looks like.

In general, the derivative tells us how much change there will be in $f(x)$ for a small change in x. Let us leave it at this and see if the definition of the derivative does us any good when it is applied to a few examples, without worrying further about the mathematical details involved.

Simple Examples

Let us try the definition of the derivative on a few simple functions. The first function is $f(x) = 3 + 4x$, which is the equation for a straight line with intercept 3 and slope 4. Let us see what we get when we use the definition of the derivative and differentiate this function. Because the graph of the function is a straight line, we know that for all values of x the tangent to a line coincides with the line itself. We know directly from the equation for the line that this line has a slope equal to 4, so that we should get the derivative $f'(x) = 4$.

When we substitute the value $x + \Delta x$ in this function, we get $f(x + \Delta x) = 3 + 4(x + \Delta x)$, and for the value x we get $f(x) = 3 + 4x$. From this we get for the derivative,

$$f'(x) = \lim \frac{f(x + \Delta x) - f(x)}{\Delta x} = \lim \frac{[3 + 4(x + \Delta x)] - [3 + 4x]}{\Delta x}$$

$$= \lim \frac{4\Delta x}{\Delta x} = \lim 4 = 4 \qquad \text{as } \Delta x \to 0 \tag{2.5}$$

The numerator first simplifies to $4\Delta x$ because the terms 3 and $4x$ cancel, with one plus and one minus term each. The denominator remains Δx, and with $4\Delta x$ in the numerator and Δx in the denominator we can cancel the Δx term, as long as Δx is different from 0. Recall that although Δx approaches 0 in the limit, it never quite reaches it, thus avoiding division by 0. The only thing then left is the constant 4, and a constant remains unchanged as Δx goes to 0. Thus, when we differentiate the function $f(x) = 3 + 4x$, we get the new function $f'(x) = 4$.

This gives us the first derivative on our list of derivatives. If we replace the numerical values of the intercept and slope of the line by the letters a and b, then we have found that

$$\text{if } f(x) = a + bx, \quad \text{then } f'(x) = b \tag{2.6}$$

An even simpler case occurs when we have the function $f(x) = a$. The function equals the constant a for all values of x, which means that the graph of the function is a horizontal line that intersects the y-axis at the value a. The slope of a horizontal line equals 0, so we should get $f'(x) = 0$ in this case. When we use the definition for the derivative we find that for the numerator we get $f(x + \Delta x) - f(x) = a - a = 0$, because the function is equal to a for all values of x. If the numerator in the fraction is already 0, then the fraction equals 0; there is no need to look at any limit. Thus, when a function equals a constant, then the derivative of this function equals 0, or

$$\text{if } f(x) = a \quad \text{then } f'(x) = 0 \tag{2.7}$$

The Derivative of a Sum of Two Functions

The function $f(x) = a + bx$ is really a sum of two functions $g(x) = a$ and $h(x) = bx$. We know that because $g(x) = a$ is a constant, we get $g'(x) = 0$. Similarly, when we differentiate the second function in the sum, we get $h'(x) = b$. Thus, for this example we get $f'(x) = g'(x) + h'(x)$, and it looks as if it may be that the derivative of a sum of two functions is the sum of the derivatives of the two functions. Let us see if this is generally true.

We have that $f(x) = g(x) + h(x)$. By substituting the sum and rearranging the terms, we get

$$\begin{aligned} f'(x) &= \lim \frac{f(x+\Delta x) - f(x)}{\Delta x} \\ &= \lim \frac{[g(x+\Delta x) + h(x+\Delta x)] - [g(x) + h(x)]}{\Delta x} \\ &= \lim \frac{[g(x+\Delta x) - g(x)] + [h(x+\Delta x) - h(x)]}{\Delta x} \\ &= \lim \left[\frac{g(x+\Delta x) - g(x)}{\Delta x} + \frac{h(x+\Delta x) - h(x)}{\Delta x} \right] \end{aligned} \tag{2.8}$$

The question now is what do to with the limit of a sum. We borrow a result from the mathematical study of limits that says that the limit of a sum is a sum of the limits. Then we get

$$f'(x) = \lim \frac{g(x+\Delta x) - g(x)}{\Delta x} + \lim \frac{h(x+\Delta x) - h(x)}{\Delta x} \tag{2.9}$$

Because the first limit is the derivative of the function $g(x)$ and the second limit is the derivative of $h(x)$, we have now shown that

$$\text{if } f(x) = g(x) + h(x) \text{ then } f'(x) = g'(x) + h'(x) \qquad (2.10)$$

That the derivative of a sum of functions is the sum of the derivatives of the functions is a useful result.

Other Functions

Derivative of $f(x) = x^2$. This is the function used as an illustration earlier. We are now ready to consider how to differentiate this function. To do this we start with the definition of the derivative, and for this particular function we get

$$f'(x) = \lim \frac{(x+\Delta x)^2 - x^2}{\Delta x} = \lim \frac{x^2 + 2x\Delta x + \Delta x^2 - x^2}{\Delta x}$$

$$= \lim(2x + \Delta x) = 2x \qquad \text{as } \Delta x \to 0 \qquad (2.11)$$

The numerator consists of the difference $f(x + \Delta x) - f(x)$, and when we square the first term we get the three first terms in the next fraction. The two terms x^2 cancel because one is positive and the other is negative. We can then divide by the term Δx, because it is different from 0 even though it is very small, and that gives us $2x + \Delta x$. As Δx gets closer and closer to 0, this sum gets closer and closer to the first term $2x$. Thus, we start with the function $f(x) = x^2$, and we find its derivative to be the function $f'(x) = 2x$.

Derivative of $f(x) = x^n$. When $n = 0$, the function becomes $f(x) = x^0 = 1$. Because the function $f(x)$ now equals a constant, the derivative of the function becomes 0. When $n = 1$ we get $f(x) = x$, which is the equation for a 45 degree line through the origin with slope equal to 1. The derivative is $f'(x) = 1$. For $n = 2$ the function becomes $f(x) = x^2$, and the derivative $f'(x) = 2x$. It may still be difficult to see the rule that applies in all three cases, but it turns out that for the general exponent n the derivative becomes nx^{n-1}. It turns out that n does not have to be an integer: The rule even works for fractions. Mathematically speaking, the rule works for any real number in the exponent.

To find the derivative of $f(x) = x^n$, let us start with the general definition of the derivative. For this particular function we get

$$f'(x) = \lim \frac{(x + \Delta x)^n - x^n}{\Delta x} \qquad (2.12)$$

First we need to multiply out and expand the term $(x + \Delta x)^n$. That gives us

$$= \lim \frac{(x^n + nx^{n-1}\Delta x + \ldots + \Delta x^n) - x^n}{\Delta x} \qquad (2.13)$$

The missing terms denoted by the three dots involve products of constants defined by n, x to some exponent between $n - 2$ and 1, and Δx to some exponent between 2 and $n - 1$.

When we open up the parentheses in the numerator, we see that the two x^n terms cancel each other out. In the next term there is a Δx in both the numerator and the denominator, so they cancel and leave us with nx^{n-1}. After canceling one Δx from each of the remaining terms, they will still all have at least one Δx as part of the product. This means that as Δx goes to 0, so do all those terms. The only thing we are left with is the term nx^{n-1}, and we get

$$\text{if } f(x) = x^n \quad \text{then } f'(x) = nx^{n-1} \qquad (2.14)$$

This means we are now able to find the derivative of any polynomial function. For example,

$$\text{if } f(x) = x^{10} + 7x^9 + 3x^6 - x^3 - x + 5$$

$$\text{then } f'(x) = 10x^9 + 63x^8 + 18x^5 - 3x^2 - 1 \qquad (2.15)$$

As a second example, suppose $n = \frac{1}{2}$. By definition, an exponent of $\frac{1}{2}$ is the same as the square root. Thus,

$$\text{if } f(x) = \sqrt{x} = x^{\frac{1}{2}}$$

$$\text{then } f'(x) = (\tfrac{1}{2})x^{\frac{1}{2}-1} = (\tfrac{1}{2})x^{-\frac{1}{2}} = \frac{1}{2x^{\frac{1}{2}}} = \frac{1}{2\sqrt{x}} \qquad (2.16)$$

Derivative of sin x. Next we consider the trigonometric function $f(x) = \sin x$. According to first principles, the derivative of this function is found from

$$f'(x) = \lim \frac{\sin(x + \Delta x) - \sin x}{\Delta x} \tag{2.17}$$

As the next step, we make use of the formula for the sine of the sum of two angles. That gives us

$$f'(x) = \lim \frac{[\sin x \cos \Delta x + \cos x \sin \Delta x] - \sin x}{\Delta x} \tag{2.18}$$

This does not look like much of a simplification, but we can factor out $\sin x$ from the first and the last term. That gives us

$$f'(x) = \lim \frac{\sin x(\cos \Delta x - 1) + \cos x \sin \Delta x}{\Delta x} \tag{2.19}$$

Because the limit of a sum is the sum of the limits, we get

$$= \lim \frac{\sin x(\cos \Delta x - 1)}{\Delta x} + \lim \frac{\cos x \sin \Delta x}{\Delta x} \tag{2.20}$$

The term $\sin x$ in the first fraction does not depend on Δx, so it acts as a constant and we can take $\sin x$ outside the limit operation. Similarly, $\cos x$ in the second term does not depend on Δx, so we can take it outside the limit operation. That gives us

$$f'(x) = \sin x \lim \frac{\cos \Delta x - 1}{\Delta x} + \cos x \lim \frac{\sin \Delta x}{\Delta x} \tag{2.21}$$

Let us look at each of these limits. In the first limit, as Δx gets close to 0, then $\cos \Delta x$ gets close to 1, and by subtracting 1 the numerator gets close to 0. Because the numerator goes to 0 faster than the denominator, the fraction has 0 as its limit. For example, measuring angles in radians and not degrees, and if the denominator $\Delta x = 0.1$, then the numerator becomes $\cos 0.1 - 1 = -0.005$, which is much smaller than the denominator 0.1. We find similar results for other small values of Δx. In the second limit, the numerator and denominator go to 0 together at about the same speed. For example, if $\Delta x = 0.1$, then $\sin 0.1 = 0.0998$; this divided by 0.1 gives us a fraction equal to 0.998. In the limit, the second fraction equals 1. The only thing left of the two terms above when Δx goes to 0 is therefore the term $\cos x$ in the second term. That means,

$$\text{if } f(x) = \sin x \quad \text{then } f'(x) = \cos x \tag{2.22}$$

Without going through the derivation we also get

$$\text{if } f(x) = \cos x \quad \text{then } f'(x) = -\sin x \tag{2.23}$$

We can also find the derivatives of the other trigonometric functions.

Differentiating the Logarithm. From time to time we run across the function $f(x) = \ln x$ for $x > 0$, where ln stands for the natural logarithm of x. Natural logarithm means we use the number $e = 2.71828\ldots$ as the base for the logarithm. When we differentiate this function we get

$$\text{if } f(x) = \ln x \quad \text{then } f'(x) = \frac{1}{x} = x^{-1} \quad x > 0 \tag{2.24}$$

From what we know about this function and from looking at the derivative, we see that for values of x close to 0 the function is very steep and rises along the negative values of y. The function crosses the x-axis at $x = 1$ and then gradually flattens out but keeps rising to infinity with increasing values of x. For very large values of x the derivative $1/x$ is almost equal to 0, meaning that the curve is almost horizontal. Here is another example of how the derivative helps us get an idea of what the graph of the original function looks like. Figure 2.4 shows a graph of the logarithmic function.

This derivative is a strange result in some ways. When we differentiate the function $f(x) = x^n$ for different values of the exponent n, then we get x to another exponent as the derivative. There exists no value of n that gives us a power function that has x^{-1} as its derivative. This is the only derivative with an integer in the exponent where the original function is not of the form x^n. Because the differentiation reduces the exponent by 1, it seems as if we would have to start with the function $f(x) = x^0$ to get an answer that involves x^{-1}, but if we use the differentiation rule for x to an exponent here,

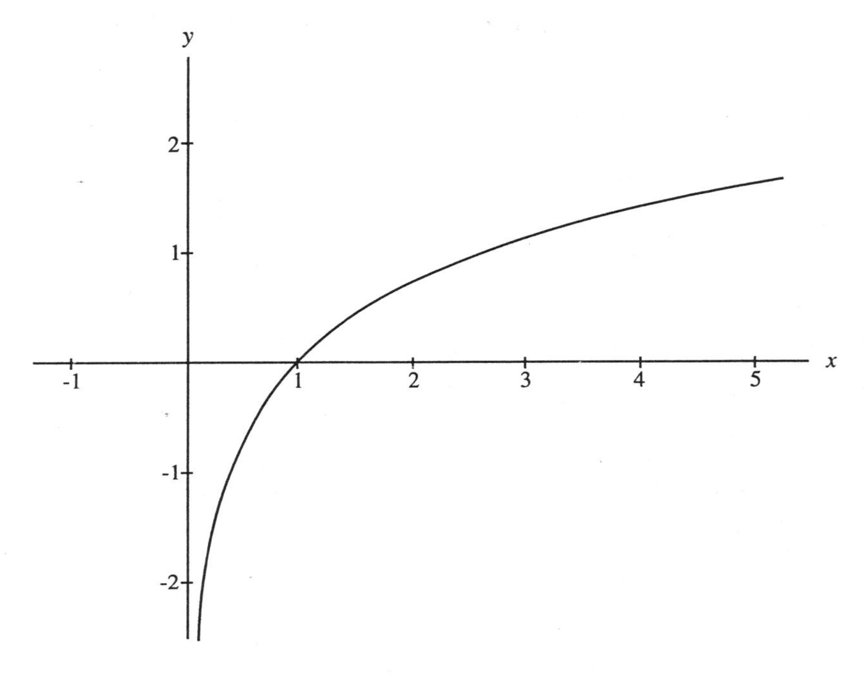

Figure 2.4. Graph of the Logarithmic Function $y = \ln x$

then we get that the derivative becomes $0(x^{-1})$. This product equals 0 because of the factor 0 in front. We also know that $x^0 = 1$, and the derivative of a constant equals 0.

Derivative of the Exponential Function. Let us consider the function $f(x) = e^x$, known as the exponential function. The number e is the same constant we saw above in the discussion of the logarithmic function. This function gets very large when x is large because x appears as an exponent of a number larger than 1. Thus, because the function gets very steep, we expect the derivative to be a function that also is large when x is large. It turns out that

$$\text{if } f(x) = e^x \quad \text{then } f'(x) = e^x \tag{2.25}$$

Thus, we have the remarkable result that for any value of x the derivative of the exponential function at that point is equal to the function itself.

This fact is the reason why the exponential function often is used to describe phenomena of growth. If we assume that the growth of a population is proportional to the size of the population, then we end up with the exponential function describing this growth. Growth (change) corresponds to the number of "offspring" at a certain time, and the larger a population is, the larger we often expect the growth to be.

Derivatives of a Product and a Quotient of Two Functions

Product. We already have found that if one function is the *sum* of two other functions, then the derivative of this sum is the sum of the derivatives of the two functions. Suppose now that we have a function that is a *product* of two other functions. That is,

$$f(x) = g(x)h(x) \tag{2.26}$$

For example, suppose we have

$$f(x) = x^2 \sin x \tag{2.27}$$

and we want the derivative of this product of the function x^2 and the function $\sin x$. It turns out that what we are looking for is *not* simply the product of the two derivatives.

Instead, we get for the derivative of a product of two functions that

$$\text{if } f(x) = g(x)h(x) \quad \text{then } f'(x) = g'(x)h(x) + g(x)h'(x) \tag{2.28}$$

In other words, the derivative of a product of two functions equals the derivative of the first function times the second function plus the first function times the derivative of the second function. For our example we get

$$f'(x) = 2x \sin x + x^2 \cos x \tag{2.29}$$

Quotient. When a function equals a quotient with one function in the numerator and another function in the denominator, we can find the derivative of this fraction the following way:

$$\text{If } \ f(x) = \frac{g(x)}{h(x)} \qquad \text{then } \ f'(x) = \frac{g'(x)h(x) - g(x)h'(x)}{h(x)^2} \tag{2.30}$$

This says that the derivative of a quotient of two functions equals the derivative of the numerator times the denominator minus the numerator times the derivative of the denominator, all divided by the square of the denominator. This is the kind of rule people who have studied calculus for a while can recite in their sleep.

As an example, suppose

$$f(x) = \frac{\sin x}{x^2} \tag{2.31}$$

According to our rule, we find

$$f'(x) = \frac{\cos x\, x^2 - \sin x\, 2x}{x^4} = \frac{x \cos x - 2 \sin x}{x^3} \tag{2.32}$$

The Chain Rule

By now we can differentiate many different functions by using the rules we have found. Sometimes, however, we get more complicated functions that do not fit any of the rules developed so far. In particular, suppose we have a function within another function. For example, we might try to find the derivative with respect to x of something like the function $f(x) = \sin x^2$.

Here we look for the derivative not of $\sin x$ itself but of the sine of a function of x, in this case the square of x. It turns out that we find the derivative of such a more complicated function by first taking the derivative of the "outside" function (sine in this example); we then multiply that by the derivative of the "inside" function (x^2 in this example). This rule is known as the *chain rule.* For the example this means we first take the derivative of the sine function, then multiply that by the derivative of the x^2 function. That gives

$$f'(x) = (\cos x^2)\,(2x) \tag{2.33}$$

Let us try this rule on a simpler example. Suppose we have $f(x) = (x^4)^2$. Let us first multiply out the exponents and find the derivative directly. Then let us use the chain rule and see if we get the same result. We know that this function is the same as $f(x) = x^8$, so the derivative becomes $f'(x) = 8x^7$. Here is a case where we can manipulate the original function so that we do not need the more complicated chain rule.

Even though we now know the correct answer, let us use the chain rule anyway to verify the result. The "outside" function is the square, and the derivative of that is 2 times the function. The "inside" function is the fourth power, and the derivative of that is 4 times the quantity cubed. Thus, we get

$$f'(x) = (2x^4)(4x^3) = (2)(4)x^{4+3} = 8x^7 \tag{2.34}$$

as before.

Formally, the chain rule can be stated,

$$\text{if } f(x) = g(h(x)) \quad \text{then } f'(x) = g'(h(x))h'(x) \tag{2.35}$$

This may look more complicated than it really is, but it says the same thing expressed in words above. In the first example, the g function is the sine and the h function is the square.

There also exists another way to describe the chain rule. When we have a function $f(x)$, we denote the derivative by $f'(x)$. We can also write $y = f(x)$ and denote the derivative as dy/dx. With this notation we can write the chain rule as

$$\frac{dy}{dx} = \frac{dy}{du} \cdot \frac{du}{dx} \tag{2.36}$$

Here the function $g(x)$ is denoted as $u = g(x)$. The rule says we first take the derivative of y with respect to u, and then we multiply by the derivative of u with respect to x. This makes it the same rule as before. We see that the equality is true because it is as if du cancels out and we are left with dy/dx on both sides of the equals sign.

Different Notations

So far we have mainly denoted a function as $f(x)$ and its derivative as $f'(x)$. Sometimes we see the derivative expressed in different ways. If $y = f(x)$ is our function, then we also can write the derivative in any of these other ways:

$$f'(x) = y' = \hat{y} = \frac{d}{dx}f(x) = \frac{dy}{dx} = D(f(x)) \tag{2.37}$$

For the simple examples we have seen so far it makes very little difference how we express the derivative, but for more complicated functions certain notations are more convenient than others in avoiding ambiguity. This often occurs when we have functions of more than one variable.

Partial Derivatives. Suppose we have a function of two variables u and v. Let

$$f(u, v) = u^2 + 3uv^2 + v^4 \tag{2.38}$$

What do we mean by the derivative of this function? First we have to specify which of the two variables we want to differentiate with respect to u or v. If we differentiate with respect to u, then v is simply a constant and the derivative of v itself equals 0. That means we get $2u + 3v^2$ as the derivative of the function. Similarly, if we differentiate with respect to v, then u is considered a constant with a derivative of 0, and the derivative of the function becomes $6uv + 4v^3$.

Because there is a choice of variables, we use the symbol ∂ for differentiation to show that this case is different from the case with only one variable. Here is an example of how we write derivatives of a function of two variables:

$$\frac{\partial}{\partial u} f(u,v) = \frac{\partial}{\partial u}(u^2 + 3uv^2 + v^4) = 2u + 3v^2 \tag{2.39}$$

$$\frac{\partial}{\partial v} f(u,v) = \frac{\partial}{\partial v}(u^2 + 3uv^2 + v^4) = 6uv + 4v^3 \tag{2.40}$$

The ∂ in the numerator says we shall differentiate something, and the ∂u and ∂v in the denominators tell us which variable it is we differentiate with respect to u or v. When we write the derivatives this way, then there is no doubt about what we have done and what variable is of interest. Another way to express the same thing is to use subscripts and write for the two derivatives

$$D_u f(u, v) \quad \text{and} \quad D_v f(u, v) \tag{2.41}$$

When we have a function of two variables, and when we differentiate such a function with respect to one of the variables, we sometimes say that we find the *partial* derivative of the function with respect to that variable.

In Chapter 4 we see that partial derivatives help us get a better understanding of the regression coefficients in multiple regression.

Higher Derivatives

When we have a function $f(x)$ and differentiate this function with respect to x, we have seen how we get another function $f'(x)$. Many times there is no reason to stop there. We can go on and differentiate the new function $f'(x)$, which gives us yet another function. This new function is known as the second derivative of the original function $f(x)$, and it can be denoted $f''(x)$. We can even go on from there and find the third and higher derivatives. Sometimes a derivative ends up being a constant, and then we can take only one more derivative before we end up with 0 and it becomes uninteresting to continue. For example, if $f(x) = x^3$, then $f'(x) = 3x^2, f''(x) = 6x$, and the third derivative equals 6, where we stop. For other functions we can keep differentiating as many times as we want. Many times we are not interested in more than the second derivative.

Maximum or Minimum. One reason we are interested in the second derivative is because of what this derivative tells us about the graph of the original function. The first derivative gives the slope of the tangent to the curve of the function for each value of the variable. If the derivative is equal to 0 for any value of the function, then we know that the tangent to the curve is horizontal for that value of the function. This means that at such a point the curve has either a minimum or a maximum (or a point of inflection). The value of the second derivative at that point can tell us whether the curve has a minimum or a maximum at that point.

A maximum or a minimum for a function at a given value of x may only mean that we have a so-called *local* maximum or minimum. In Figure 2.6 we see that the function has a maximum value at $x = a$, because the values of the function fall off on both sides of a. It is only a local maximum in the sense that the function has even larger values than at $x = a$ as x increases beyond the value b out toward positive infinity.

Suppose we have the function $f(x) = x^2 - 8x + 20$. Where do we have a minimum or a maximum value of this function? In other words, where is the first derivative equal to 0? We differentiate the function and set the resulting function equal to 0. Solving for x and setting the derivative equal to 0 gives us $f'(x) = 2x - 8 = 0$. This is true when $x = 4$.

When x is less than 4 we see that the derivative is negative; in other words, the function comes down with increasing values of x. When x is

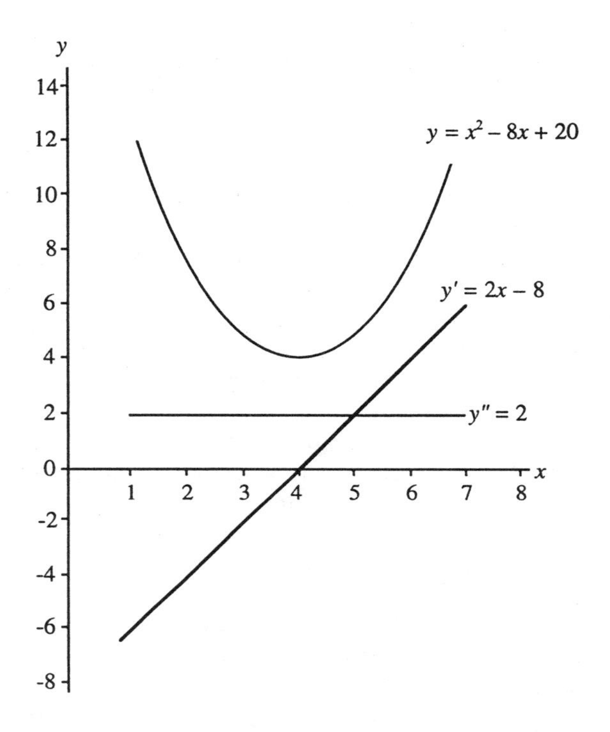

Figure 2.5. A Parabola With Its Derivative (Slanted Line y') and Second Derivative (Horizontal Line y'')

larger than 4 we see that the derivative is positive. This tells us that the function goes up with increasing values of x. This tells us that at $x = 4$, where the slope of the tangent is 0, the function must have a minimum.

Figure 2.5 shows our function, its first derivative, and its second derivative. The function itself is ∪-shaped and is known as a parabola. We see that as x increases from the left, the function comes down and bottoms out at $x = 4$. After that the function increases again. This means that the derivative of the function should be negative for values of x less than 4, because any tangent to the curve will have a negative slope when x is less than 4. Similarly, the tangents have positive slopes for x greater than 4, so the derivative should be positive for x larger than 4. The derivative is shown by the straight line marked $y' = 2x - 8$. We see that for x less than 4, the

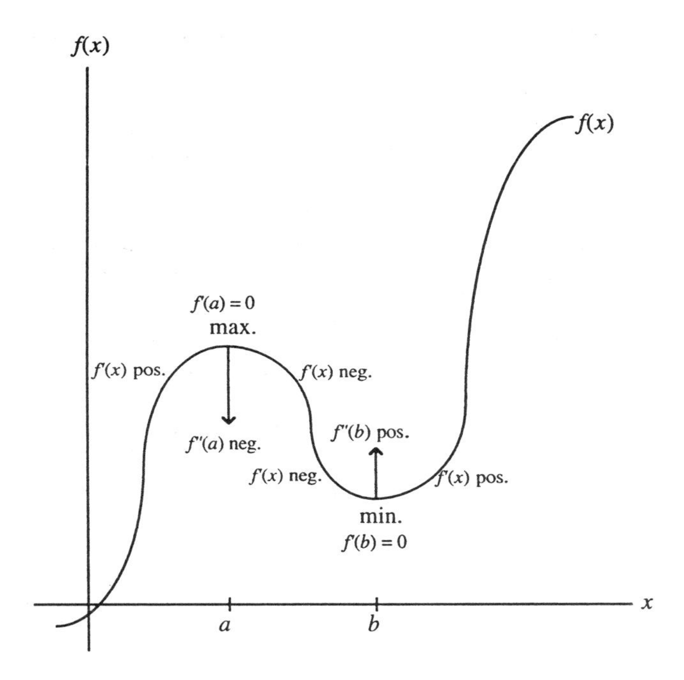

Figure 2.6. Values of First and Second Derivatives for a Function With a Minimum

line has negative y values, and for x larger than 4 the line has positive y values, as we expected.

We also see that when x is very different from 4, the curve becomes very steep and the slope of the tangent is therefore very large in absolute value. When x is very different from 4, then the corresponding points on the line for $f'(x)$ will be large, as they should be because the tangents have large slopes.

Turning to the line for $f'(x)$ we see that the slope for this line equals 2 for all values of x. This is the slope we get when we differentiate $f'(x)$ and get $f''(x)$. In the graph we see this represented as a horizontal line for the second derivative.

The second derivative is helpful for determining whether we have a minimum or a maximum. Here we find $f''(x) = 2$, which is positive for all values of x. Here the first derivative is a straight line with positive slope 2.

That means that for small enough values of x, the line is below the x-axis and has negative values. For large enough values of x, the line is above the x-axis and has positive values. In other words, the slope of the tangent to the curve is first negative and then positive. This means here the curve must have a minimum. Any time we have a *positive* second derivative at a point where the first derivative is equal to 0, we have a *minimum* for the function. Similarly, any time we have a *negative* second derivative at a point where the first derivative is equal to 0, then we have a *maximum* for the function.

These results are summarized in Figure 2.6 for the general case. The arrow pointing up at the local minimum of the function at $x = b$ is to remind us that there the second derivative is positive. The arrow pointing down at the local maximum of the function at $x = a$ is to remind us that there the second derivative is negative.

As a final point, note in the figure that at $x = a$ the function has a maximum, but this maximum value of the function is only a local maximum. This is because we can find other values of the variable where the value of the function is even larger than at the maximum we have identified. If we choose large values of x, we see that the values of the function itself are larger than the value at the local maximum point. Similarly, we can find small values of x where the value of the function is smaller than the value at the local minimum point. We cannot tell from the values of the first and second derivatives only, for a given x, whether the extreme point is a local extreme point or not.

List of Derivatives

if $f(x) = a$	then $f'(x) = 0$
if $f(x) = a + bx$	then $f'(x) = b$
if $f(x) = x^n$	then $f'(x) = nx^{n-1}$
if $f(x) = \sin x$	then $f'(x) = \cos x$
if $f(x) = \cos x$	then $f'(x) = -\sin x$
if $f(x) = \ln x$	then $f'(x) = \frac{1}{x} = x^{-1}$

$$\text{if } f(x) = e^x \qquad \text{then } f'(x) = e^x$$

$$\text{if } f(x) = g(x) + h(x) \qquad \text{then } f'(x) = g'(x) + h'(x)$$

$$\text{if } f(x) = g(x)h(x) \qquad \text{then } f'(x) = g'(x)h(x) + g(x)h'(x)$$

$$\text{if } f(x) = \frac{g(x)}{h(x)} \qquad \text{then } f'(x) = \frac{g'(x)h(x) - g(x)h'(x)}{h(x)^2}$$

$$\text{if } f(x) = g(h(x)) \qquad \text{then } f'(x) = g'(h(x))h'(x)$$

3. INTEGRATION

Introduction

Integration makes up the second part of calculus. This part is sometimes known as integral calculus, to distinguish it from the differential calculus in the previous chapter.

When we integrate a mathematical function, we get another mathematical function. We assume throughout that the function we start with is continuous, such that it is well behaved and does not have any jumps. All the functions we consider here satisfy these criteria and behave such that they can be integrated.

The function we start with is often denoted $f(x)$, and the function we get after integration is often denoted $F(x)$. For example, by integrating $f(x)$ we get

$$\text{if } f(x) = 2x \qquad \text{then } F(x) = x^2 + constant$$

$$\text{if } f(x) = \cos x \qquad \text{then } F(x) = \sin x + constant. \qquad (3.2)$$

When we do integration this way, the answer function always includes a constant. The value of this constant is not determined by the integration method itself, and the function $F(x)$ is therefore not uniquely determined. It can be one of many different functions, depending upon the value of the constant. This integration constant often plays a minor role, however, and as we see later in this chapter, many times we need not pay any attention to this constant, and we need not focus on this issue here.

Sometimes we carry integration one step further. After we do the integration and find the function $F(x)$, we go on to say that we want to integrate the original function $f(x)$ from one value of the variable x to another value. The result of this operation is a numerical value. For example, suppose we want to integrate the function $f(x) = 2x$ from 1 to 2. The answer we get in this case is 3.

One part of learning calculus consists of learning the rules for integration. The rules have a theoretical underpinning, but this usually does not help when we want to integrate a particular function. Thus, the only way to do the integration is to know the rules, unless we can find our integral on a list of integrals or we have a computer program that can help find the integral for us, or at least make a good guess.

The reason integration becomes so important is that it answers questions about phenomena in the world that we can express mathematically. For example, when in statistics we compute a value of the t variable and want to find the corresponding p value, we perform an integration. (The t is a statistical variable often used in hypothesis testing to determine whether a null hypothesis should be rejected. The p value is the probability that our test statistic equals the observed value or more extreme values, given that the null hypothesis is true.) It so happens that in this case the computations that went into the integration already have been made for us, and we can look up at least an approximate numerical answer in a statistical t table.

The Definite Integral

Definition. Look at the function in Figure 3.1. Suppose we want to find the area under the smooth curve displayed by the function $f(x)$ between the two values a and b of the variable x. On the face of it, this may not be a very interesting question, but it turns out to lie at the heart of integration and has meanings beyond the way we have phrased the question so far.

If the function had been a straight line instead of a curve, then the area would have been easy to find. Indeed, if the curve had been a straight horizontal line, then the area would simply be the area of a rectangle, which we could find by multiplying the base by the height of the rectangle. In our case, it is more difficult to find the area we want because the curve in Figure 3.1 is not a straight line. If we cannot find the area directly, maybe we can find an approximate answer. This is what the figure shows, and this is where integration enters.

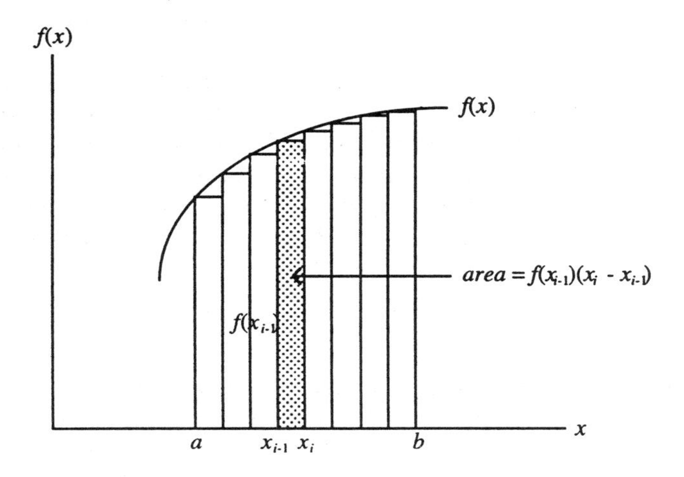

Figure 3.1. Area Under a Curve Between $x = a$ and $x = b$

First we introduce the values $x_1, x_2, \ldots, x_{n-1}$ of the variable x between a and b. This divides the line from a to b into n intervals. We do this in such a way that the length of each of these intervals is less than some small number δ. The i-th interval now goes from x_{i-1} to x_i. At the point x_{i-1}, the value of the function equals $f(x_{i-1})$. The area of the rectangle with base $x_i - x_{i-1}$ and height $f(x_{i-1})$ then becomes the product $f(x_{i-1})(x_i - x_{i-1})$. In Figure 3.1 the area we want has been replaced by the combined areas of several thin rectangles.

If we add up the areas of these thin rectangles, we get something that approximates the entire area under the curve. The sum of these rectangles will be somewhat smaller than the area under the curve because we will be missing the small, almost triangularly shaped areas on the tops of the rectangles. If we make the length of the intervals small, however, then we will not be missing very much. In other words, we want the sum of these areas as the limit when δ goes to 0, because, by definition, all the intervals are shorter than δ.

Formally, we now have that the area A we want under the curve can be written as

$$\lim_{\delta \to 0} [f(a)(x_1 - a) + f(x_1)(x_2 - x_1) + \ldots + f(x_{n-1})(b - x_{n-1})] = A \qquad (3.3)$$

We did not need to present this discussion in terms of areas of rectangles because the statement above is a general statement about mathematical functions. There are times when this limit is not in terms of areas, but the limit becomes more concrete if we look at in terms of areas. In addition, it turns out that we get the same value A for this limit if we use any point within each interval to find the value of the function and not necessarily use the left endpoint. The notation gets simpler, however, if we do it the way it is done here. The value A depends only on the two endpoints a and b and what function $f(x)$ we have. Needless to say, this is a very powerful mathematical result we have for A. The presentation here does not contain a proof of this result.

The next step consists of simplifying the expression for the area A. The expression we have for A is rather cumbersome, and we make use of this expression so often that we would like to be able to write it differently. First we introduce the summation sign Σ and write the expression as

$$\lim_{\delta \to 0} \sum_{i=1}^{n} f(x_{i-1})(x_i - x_{i-1}) = A \tag{3.4}$$

When $i = 1$ we get the value x_0, which we have called a. Similarly, when $i = n$ we get the value x_n, which we have called b. Next we replace all the widths of the intervals by the term dx and write the limit of the sum in this shortened way:

$$\lim \sum_{a}^{b} f(x)dx = A \tag{3.5}$$

This can be interpreted as saying that we divide the interval from a to b into small intervals, and the width of each is denoted dx. We multiply the value of the function at some value within the small interval by dx and add all these products. We are still summing over i, but that amounts to finding the sum from the smallest value of x at a to the largest value at b, so we write the sum as if we are summing from a to b. This sum approaches in the limit the value A when widths of all the small intervals go to 0.

Finally we pull the limit sign and the summation sign together and write the limit as

$$\int_a^b f(x)dx = A \tag{3.6}$$

The limit sign and the summation sign have now been replaced by single symbol that looks like a slanted S. This is the integration sign. It looks like a slanted S because it is supposed to remind us that integration means a sum of many small terms that in the limit gives us what we want. We read this as saying the limit of the sum of the products $f(x)dx$ from $x = a$ to $x = b$. Such a limit does not always exist, but when the function $f(x)$ is reasonably well behaved, we can find the limit. For all the functions we look at in this book, the limits do exist for reasonable choices of the two boundaries a and b.

We call the expression above a *definite* integral, with lower limit a and upper limit b. Below we also discuss indefinite integrals. A definite integral gives us a number, and this number equals the size of the area defined by the function $f(x)$, the x-axis, and the two boundaries a and b. From the way the definite integral is defined, we get that areas below the x-axis are negative. Thus, if the function crosses the x-axis somewhere between a and b, the total area we are interested in may consist of a positive area above the x-axis and a negative area below the x-axis in such a way that the total area, when found directly from an integral, may even become equal to 0 if the positive area is equal to the negative area.

As an example, when we do the following integration we get the value 3. (The actual integration is done in some detail below.)

$$\int_1^2 2xdx = 3 \tag{3.7}$$

This tells us that the area under the curve with the equation $y = 2x$ between $x = 1$ and $x = 2$ equals 3 in the limit when the rectangles introduced above get very thin. We already knew 3 to be the answer. This is because the figure we consider is a trapezoid with base 1 and one height equal to 2 and the other equal to 4, as seen in Figure 3.2. From the figure we see directly that the trapezoid has an area equal to 3, because we find the area of a trapezoid as one half of the sum of the two parallel sides times the width. Here we get $(\frac{1}{2})(2 + 4)1 = 3$.

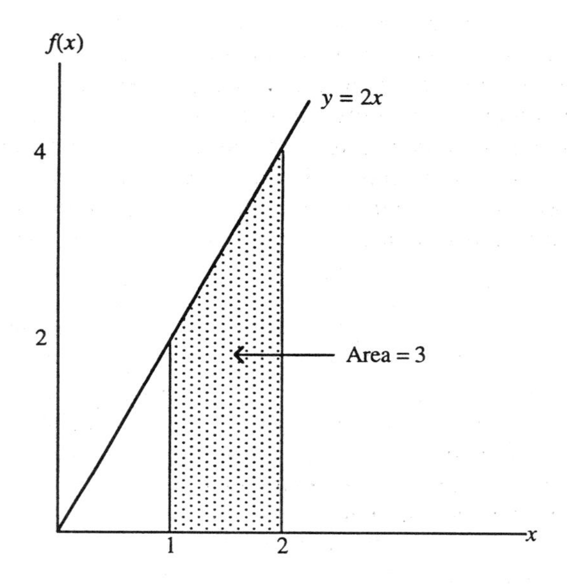

Figure 3.2. Area Under the Curve of the Function $y = 2x$ Between $x = 1$ and $x = 2$

Area and Slope

Although we can use integration to find areas under mathematical curves, there is a more fundamental idea that underlies integration. The more basic issue in integration is how we start with the function $f(x)$ and integrate to find a function $F(x)$. In symbols we have

$$F(x) = \int f(x)dx \tag{3.8}$$

In words, $F(x)$ is the function we get when we integrate the function $f(x)$. Here there is no reference to areas or anything else. The question is simply how we go from one function $f(x)$ and apply the rules of integration to find another function $F(x)$. $F(x)$ cannot be just any function: It has to be the function that gives us the proper area when we phrase the question in terms of areas again. This integral, without the mention of any boundaries a and b, is called an *indefinite* integral, as opposed to the definite integrals discussed earlier.

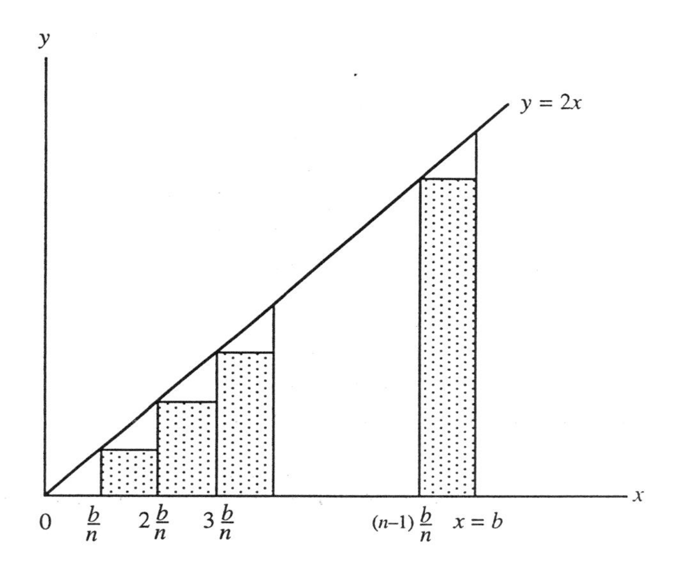

Figure 3.3. Adding Rectangles

The remarkable answer to this question is that $F(x)$ is a function that has the function $f(x)$ as its derivative. For example, when we integrate $f(x) = 2x$, this is why we get

$$F(x) = x^2 + c \quad \text{from} \quad \int 2x\, dx \tag{3.9}$$

where c is some unknown constant. It is because we are not able to determine the value of the constant that we call this an indefinite integral. When we now differentiate $F(x) = x^2 + c$, we get $F'(x) = 2x$ (the derivative of the constant c equals 0 no matter what it is equal to). We now recognize $f(x) = 2x$ as the original function under the integration sign.

Let us look at this point in greater detail by returning to the question of area. Figure 3.3 shows the function $f(x) = 2x$ from the lower value of $x = 0$ to the upper value $x = b$. We want to find the area under the curve across this range of values. From the figure we see that the area we want is a right triangle with base b and height $2b$. The height is $2b$ because we are at the place where $x = b$, and when we substitute this value of x into the function

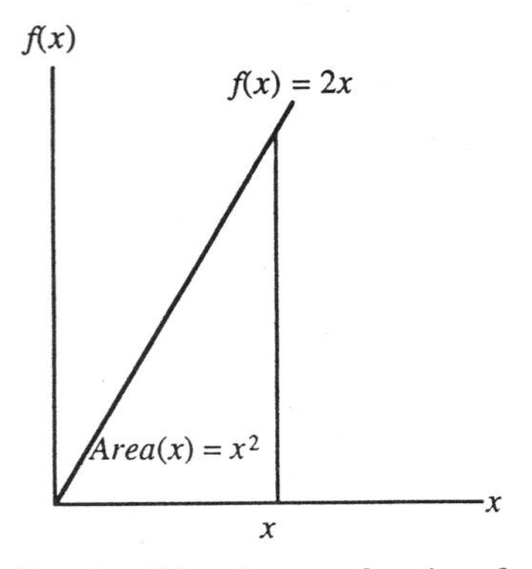

Function $f(x)$ and area as function of x.

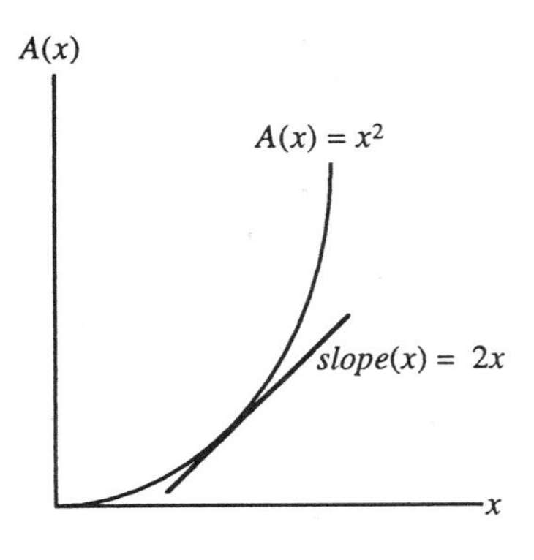

Area function with slope as function of x.

Figure 3.4. Area as a Function of x and Slope as a Function of Area

$f(x) = 2x$ we get $2b$. The area of a triangle equals one half times the base times the height, or

$$Area = \tfrac{1}{2}(b)(2b) = b^2 \tag{3.10}$$

Next, suppose we look not at a specific value b but at a general value x of the x variable. Thus, the base of the triangle goes from 0 to x, the height of the triangle becomes $2x$, and the area therefore becomes x^2. The actual magnitude of the area depends on what value we choose for x, so we can write the area as a function of x, $A(x) = x^2$.

We started with the original function $y = 2x$, and we ended up with the area under the curve becoming $A(x) = x^2$. Looking back at the previous chapter on differentiation, we see that $A(x) = x^2$ is a function that has $f(x) = 2x$ as its derivative. This is no accident, and this fact applies to functions other than $f(x) = 2x$ as well, even though no proof is provided for this.

Figure 3.4 summarizes the connection between area and slope. We start with a certain function $f(x)$ defined across a range of values of the variable, here from 0 to positive infinity. The area under the curve from 0 to x becomes a function of x. We can then graph the area function. When we differentiate the area function and find the slope of the tangent line as a function of x, we get back the original function $f(x)$. This is only an example for one specific function, but the example illustrates the connection between area and the slope of the tangent line.

Adding Small Rectangles

Let us also convince ourselves that we get the same area when we use the basic underlying idea of integration and add up all the areas of all the rectangles in Figure 3.3. We divide the line from 0 to b into n equal pieces, such that the length of each piece becomes b/n. On top of each piece we then construct a rectangle. Thus, the base of each rectangle equals b/n. Because we have the function $y = 2x$, we get the height of the first rectangle when we substitute the value $x = b/n$ in this function. The height becomes $2(b/n)$. Similarly, the height of the second rectangle becomes $2(2b/n)$, the height of the third rectangle becomes $2(3b/n)$, and so on, up to the height of the last rectangle, which becomes $2(n - 1)b/n$. When we multiply the base with length b/n by the height of each rectangle and add these products, we get the combined area of all the shaded rectangles in Figure 3.3.

This total area of all the rectangles becomes

$$\frac{b}{n}2\frac{b}{n}+\frac{b}{n}2\frac{2b}{n}+\frac{b}{n}2\frac{3b}{n}+\ldots+\frac{b}{n}2\frac{(n-1)b}{n} \tag{3.11}$$

In this sum we can factor out $2\left(\frac{b}{n}\right)^2 n$ from every term. That way the sum becomes

$$=2\left(\frac{b}{n}\right)^2[1+2+3+\ldots+(n-1)] \tag{3.12}$$

To find this sum we have to add up the first $n - 1$ integers. There exists a formula for such a sum. The formula tells us that this sum equals $(n - 1)n/2$, and when we substitute this expression for the sum into the expression above, we get

$$=2\left(\frac{b}{n}\right)^2\frac{(n-1)n}{2}=b^2\frac{n-1}{n}=b^2\left(1-\frac{1}{n}\right) \tag{3.13}$$

This is because we can cancel the 2s and one n in the numerator against an n in the denominator.

Because we have added up the areas of the shaded rectangles in Figure 3.3, we know we have found an area smaller than the area under the curve. This error comes from missing each of the little triangles on top of the rectangles. The smaller we make the base of each rectangle, however, the smaller the triangles become and the smaller the error becomes. We make the base of each rectangle small by having many rectangles, and we get many rectangles by choosing a large value for the number n.

Mathematically, we now ask what happens to the limit of the quantity

$$b^2\left(1 - \frac{1}{n}\right) \tag{3.14}$$

as n gets large and goes to infinity. We can see directly that in this case the fraction $1/n$ gets smaller and goes to 0 with increasing n, and the whole quantity therefore goes to b^2. We know already that this is the correct value for the area under the curve of the function $f(x) = 2x$ from 0 to b, but now we also have shown this area as the limit of the sum of the areas of many rectangles as the number of rectangles increases.

Archimedes was one of the pioneers working on such a way to find areas under curves, and this method still carries his name. The reason he is not credited with the invention of calculus is that the method only works when there exists a formula for the sum we need to compute. In the example, the function for the curve $f(x) = 2x$ is so simple that the sum we need is simply the sum of the first $n - 1$ integers. We could have looked at the more complicated function $f(x) = x^2$, and the corresponding sum for that function calls for the sum of the $n - 1$ first squared integers. That sum is not as simple to find, but the sum has a well-known formula we could use. For other functions the required sum may become too difficult to find.

What Archimedes did not realize was that the answer is always of a form such that the derivative of the answer gives back the original equation for the curve we work with. The major reason he did not realize this is probably because the derivative had not been invented yet, either by him or anyone else!

Because of the work by Newton and by Leibniz, we no longer have to find complicated sums to find the area under a curve. Now we simply look for the function that has the original $f(x)$ as its derivative. The only problem is that there sometimes do exist functions $f(x)$ so complicated that we cannot find the corresponding function that has $f(x)$ as its derivative. Then it is back to Archimedes's method again, using a computer program that can find the required sum of the areas of the small rectangles.

The Fundamental Theorem of Calculus

Let us carry this discussion one step further. Suppose we want to find the area under the curve $f(x) = 2x$ from 1 to 2, instead of from 0 to b as we did above. Figure 3.2 illustrates this area. Because of the simple shape of the area, we already have found that the answer equals 3. Now let us see if we can find this area directly, using integration.

We have $f(x) = 2x$, and we have found by integrating this function that the area from 0 to an arbitrary value x equals $F(x) = x^2$. That means that the area from 0 to $x = 2$ equals $F(2) = 2^2 = 4$, and the area from 0 to $x = 1$ equals $F(1) = 1^2 = 1$. If we subtract the area from 0 to 1 from the area from 0 to 2, then we get the difference as the area from 1 to 2. For this example, the area from 1 to 2 gives us $F(2) - F(1) = 2^2 - 1^2 = 4 - 1 = 3$, as before.

We can summarize all of this in one neat mathematical formula. We get

$$\int_1^2 f(x)dx = \int_1^2 2xdx = F(2) - F(1) = 2^2 - 1^2 = 4 - 1 = 3 \qquad (3.15)$$

where $F(x) = x^2$. Here is a case where it is not necessary to include the constant c, because it will cancel out in the difference $F(2) - F(1)$. For all practical purposes we can assume here that the constant equals 0.

The formula says that we want the area under the curve $f(x) = 2x$ from $x = 1$ to $x = 2$. We find this area by finding the function $F(x)$ that has $f(x) = 2x$ as its derivative, and that means $F(x) = x^2$. Then we substitute the upper limit $x = 2$ into $F(x)$ and from that we subtract what we get when we substitute the lower limit $x = 1$ into $F(x)$.

Finally, in the general case, when we want to integrate the function $f(x)$ from the lower limit $x = a$ to the upper limit $x = b$, then we find this area by first finding $F(x)$. That is, we need a function $F(x)$ such that $F'(x) = f(x)$. The area we want is then equal to the difference $F(b) - F(a)$. In symbols we can say the same by writing

$$\int_a^b f(x)dx = F(b) - F(a) \qquad (3.16)$$

This equality is known as the *fundamental theorem of calculus*. So far we have shown only that this works for a very simple example, but it is also true in the general case. We assume here that $f(x)$ is a well-behaved

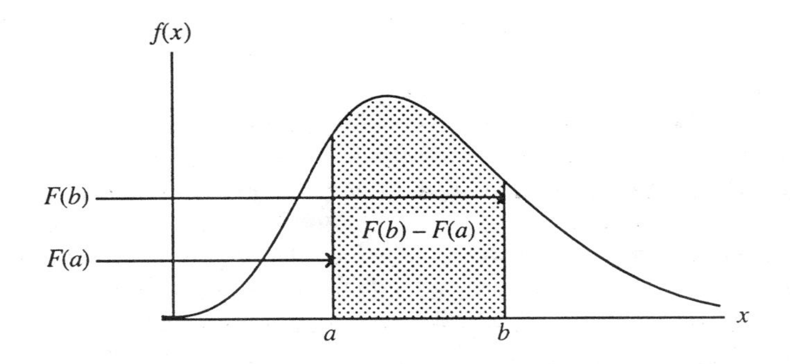

Figure 3.5. The Area Under the Curve $f(x)$ From a to b as $F(b) - F(a)$

function over the interval from a to b. Figure 3.5 shows the area we want as the difference between two areas.

We now see that integration often is a two-step process. When we are given the equation for the curve $f(x)$ and the two limits of integration a and b, we first have to find the other function $F(x)$ such that $F'(x) = f(x)$. This is not always possible to do, and there may not exist such a function for all values of x. When there is such a function, $F(x)$ is sometimes known as the anti-derivative of $f(x)$. The second thing we have to do is to substitute the two limits a and b and find the difference $F(b) - F(a)$.

In the first step there is no mention of the limits a and b; all we need to do is to find $F(x)$. This is expressed in symbols as

$$F(x) = \int f(x)dx \tag{3.17}$$

An integral like this without limits and only involving the two functions $F(x)$ and $f(x)$ is known as an indefinite integral. It has this name because we are not able to determine the value of the constant that is a part of $F(x)$.

When we include the limits a and b to find an area A, we can write this in symbols as

$$A = \int_a^b f(x)dx = F(b) - F(a) \tag{3.18}$$

This is the integral known as a definite integral when it includes the two limits of integration a and b. Thus, an indefinite integral produces another function, whereas a definite integral uses that function to find a specific numerical value.

Example. We have the function $f(x) = 3x^2 + 6x - 7$, and we want to find the area under this curve from $x = 1$ to $x = 3$. Any time we want to find an area using integration we have to remember that integration treats areas above the x-axis as positive and areas below the x-axis as negative. If part of a total area lies below and another part above the x-axis we could get a net area as small as 0 from the integration if we do not look at the two areas separately. Here we do not run into this problem because the function has only positive values across the range from 1 to 3, and the entire area therefore lies above the x-axis.

First we have to work on the indefinite integral

$$F(x) = \int (3x^2 + 6x - 7)dx \tag{3.19}$$

Even without much experience with integration we can see directly that the function we are looking for becomes

$$F(x) = x^3 + 3x^2 - 7x + c \tag{3.20}$$

This is the function we want, because if we differentiate $F(x)$ we get

$$F'(x) = 3x^2 + 6x - 7 \tag{3.21}$$

which is the original function under the integral sign. Because we can differentiate a sum by differentiating each term, we can also integrate a sum by finding the integral of each term.

Now we can find the area we want. We get

$$\begin{aligned} F(3) - F(1) &= [(3^3) + 3(3^2) - 7(3) + c] - [(1^3) + 3(1^2) - 7(1) + c] \\ &= [33 + c] - [-3 + c] = 36 \end{aligned} \tag{3.22}$$

We notice that the constant c disappears here, and this is always the case when we have a definite integral. We therefore leave the constant out when we have a definite integral.

Most of the time we do the two steps together and write them the following way.

$$Area = \int_{1}^{3} (3x^2 + 6x - 7)dx$$

$$= (x^3 + 3x^2 - 7)\Big|_{1}^{3}$$

$$= [(3^3) + 3(3^2) - 7(3)] - [(1^3) + 3(1^2) - 7(1)]$$

$$= 33 - (-3) = 36 \qquad (3.23)$$

We recognize that the second line contains the function $F(x)$ we found as the anti-derivative of the original function $f(x)$. This is followed by a vertical line with the upper limit 3 and the lower limit 1. The vertical line tells us that we should first substitute the upper limit into the function and then subtract whatever we get when we substitute the lower limit into the function.

Integration is very much like a backward procedure compared to differentiation. The rules for differentiation can be learned with practice, but when we come to integration, the situation is such that from a differentiation point of view we know the answer and we want to find the question. That is not always easy. After a while we recognize some of the more commonly occurring integrals, and there are a few methods that can help us solve integrals. There also exist lists of integrals, and they can be helpful.

The Integral as a Function of a Limit

The material above can be used to look at a special case of integration. We have seen what happens with an integral when we integrate from a lower limit a to an upper limit b. What happens if the lower limit a is at minus infinity and the upper limit b is some general value x of the variable? Figure 3.6 shows a function $f(x)$, and the area we want is shaded from minus infinity up to a general value x of the variable.

That means we are looking at the following integral:

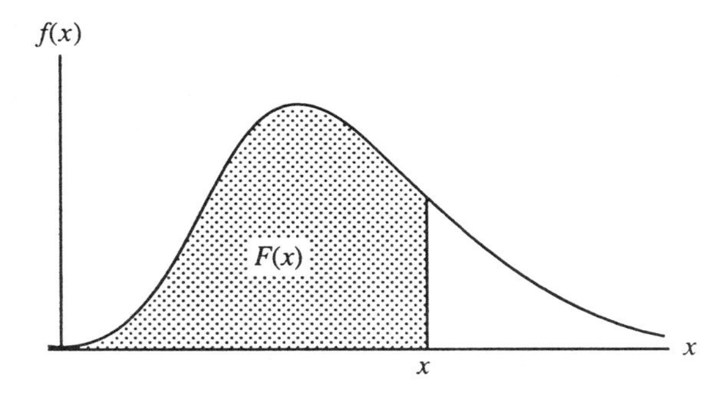

Figure 3.6. The Area $F(x)$ Under the Curve $f(x)$ From Minus Infinity to x

$$\int_{-\infty}^{x} f(x)dx = F(x) - F(-\infty) \tag{3.24}$$

The notation now becomes cumbersome. As the integral is written here, we have to keep in mind that we use the symbol x in two different ways. As the upper limit of the integration it stands for some value of the variable. As a variable in $f(x)$ it stands for the variable used in the integration. This double meaning can become confusing, and we usually write such an integral slightly differently. We recall that we have the same function whether we use x or some other symbol for the variable. Let us therefore replace the variable with the letter t. That way the integral becomes

$$\int_{-\infty}^{x} f(t)dt = F(x) - F(-\infty) \tag{3.25}$$

Just as $F(a)$ is the area under the curve to the left of $x = a$, so is $F(-\infty)$ the area under the curve to the left of minus infinity. That area is 0 because the curve does not go beyond minus infinity under any circumstance. What we then get is that the familiar function $F(x)$ can also be written

$$F(x) = \int_{-\infty}^{x} f(t)dt \quad \text{where} \quad F'(x) = f(x) \tag{3.26}$$

These are two equivalent ways we have of tying together the two functions $f(x)$ and $F(x)$. First, $f(x)$ is the function we get when we differentiate $F(x)$; second, $F(x)$ is the function we get when we integrate $f(x)$ from minus infinity to x.

Specific Integrals

Let us look at some integrals.

Polynomial Functions. Suppose $f(x) = ax^n$ $(n \neq -1)$. When we integrate this function we get

$$F(x) = \int ax^n dx = \frac{a}{n+1} x^{n+1} + c \tag{3.27}$$

We can see that this is the correct answer, because if we differentiate the right side we know that the answer is the exponent multiplied by x to the exponent minus 1. Thus, the $(n + 1)$ terms will cancel and we are left with the constant a in the front while the exponent is reduced to n, or ax^n. The derivative of the constant c is as usual equal to 0.

1/x. With the polynomial term above we had to exclude the possibility that $n = -1$. The reason for that is that the answer involves the fraction $\frac{a}{n+1}$ and we cannot have 0 in the denominator, and that is exactly what happens if $n = -1$. What happens in the case when $n = -1$? For the sake of simplicity and without any loss, let the constant $a = 1$.

For $n = -1$ we are looking at the integral

$$\int \frac{1}{x} dx \tag{3.28}$$

When we go back and look at our list of how we differentiate various functions, we find that $1/x$ involves the logarithmic function. That is, when we differentiate the logarithmic function $f(x) = \ln x$, we get $f'(x) = 1/x$. Therefore, when integrating we get

$$\int \frac{1}{x} dx = \ln x + c \tag{3.29}$$

Because we cannot take the logarithm of a negative number, we will assume that x is positive.

Now we are in the strange situation that for any value of n other than -1 we get the power solution to the integral, and for $n = -1$ we get the logarithmic answer. Even if $n = -0.999$ or -1.0001 or any other number close to -1, we still get the power answer. At exactly -1 the answer switches over to the logarithmic answer. This means that the logarithmic function $f(x) = \ln x$ fills in a gap in the power functions.

The Sine and Cosine Functions. When we integrate the sine function we get

$$F(x) = \int \sin x \, dx = -\cos x + c \qquad (3.30)$$

This is the correct answer, because we know that if we differentiate the function $F(x) = -\cos x + c$ we get $F'(x) = \sin x + 0 = \sin x$. Similarly, for the cosine function we get

$$\int \cos x \, dx = \sin x + c \qquad (3.31)$$

The Exponential Function. Finally, from the rules of differentiation we know that for the integral of the exponential function we get

$$\int e^x dx = e^x + c \qquad (3.32)$$

Integral of a Sum. If we have a function $f(x)$ that is a sum of two other functions $g(x)$ and $h(x)$, then the integral of $f(x)$ is the sum of the integrals of $g(x)$ and $h(x)$. That is,

$$\int f(x)dx = \int [g(x) + h(x)]dx = \int g(x)dx + \int h(x)dx \qquad (3.33)$$

The same rule holds for the difference between two functions. The integral of a product of two functions, however, does *not* equal the product of the integrals of the two functions. Nor does the integral of the quotient of two functions equal the quotient of the integrals of the two functions.

Sine Function. How large is the area under the sine curve from 0 degrees to 180 degrees? Over that range the curve looks like a half moon, starting at 0 for 0 degrees, going up to 1 at 90 degrees, and coming down to 0 again at 180 degrees. We get

$$\int_0^{180} \sin x dx = -\cos x \Big|_0^{180}$$

$$= -\cos 180 - (-\cos 0) = -(-1) - (-1) = 2 \qquad (3.34)$$

Note that it is more common in calculus to use radians and not degrees to measure angles. Because the reader might be more familiar with degrees, we did this simple example using degrees because it also works for degrees.

Area Under a Parabola. What is the area under the parabola with equation $y = x^2$ from 0 to 3? The curve starts at 0 and curves upward for increasing values of x. We get

$$\int_0^3 x^2 dx = \frac{1}{3} x^3 \Big|_0^3 = \frac{1}{3} 3^3 - \frac{1}{3} 0^3 = \frac{1}{3} 27 = 9 \qquad (3.35)$$

Integration Methods

We can solve many integrals simply by looking at the function under the integral sign and recognizing this function as the derivative of another function. We are usually able to do this for simple polynomials and for logarithmic and exponential functions, as well as for simple trigonometric functions. Beyond that, integration can become difficult for more complicated functions. There are published lists of integrals in elementary textbooks as well as in mathematical handbooks, and we can often find integrals we are looking for in such lists. Also, computer software exists that can do integration.

A couple of integration methods are commonly used. The name of one method is integration by substitution, and an example is discussed below. The name of the other method is integration by parts, and that method is not discussed.

Method of Substitution. We want to find the area under the curve of the function $f(x) = (1 + 2x)^2$ between 0 and 1. This means we want to find the integral

$$\int_0^1 (1 + 2x)^2 \, dx \tag{3.36}$$

We cannot do this integral directly, but one way to find the integral is to square the expression in the parentheses and get the polynomial $1 + 4x + 4x^2$. We know how to find the integral of this function because we can integrate each term separately. We get

$$\int_0^1 (1 + 4x + 4x^2)dx = (x + 2x^2 + \frac{4}{3}x^3)\Big|_0^1$$

$$= (1 + 2 + \frac{4}{3}) - 0 = 4\frac{1}{3} \tag{3.37}$$

Now that we know the answer, we can illustrate the use of the method of substitution with this integral. The integral involves an expression to the second power, and we know how to integrate the square of a variable. The only trouble is that we do not have a single variable squared; we have the whole expression $(1 + 2x)$ squared. If we write

$$y = 1 + 2x, \tag{3.38}$$

then we get the integral of y^2, and that is an integral we know how to find.

We still have the term dx, meaning that we should integrate with respect to x. To find the integral we need to replace the dx by dy. To do this we first solve the equation above for x and find

$$x = -\frac{1}{2} + \frac{1}{2}y \tag{3.39}$$

This is now a function of y, and we can differentiate this function with respect to y. Using the d-notation we get

$$\frac{dx}{dy} = \frac{1}{2} \tag{3.40}$$

We can solve this expression for dx in terms of dy. That gives us

$$dx = \frac{1}{2}\,dy \tag{3.41}$$

We can now substitute for both x and dx. We also need to be concerned, however, about the two limits of integration. We want to integrate the original function from $x = 0$ to $x = 1$. When $x = 0$ we get $y = 1 + 2(0) = 1$, and when $x = 1$ we get $y = 1 + 2(1) = 3$. This means that for y we have to integrate from 1 to 3.

Now we are ready to substitute for $1 + 2x$, dx, and the limits. We get

$$\int_0^1 (1 + 2x)^2 dx = \int_1^3 y^2 \frac{1}{2} dy = \frac{1}{3} y^3 \Big|_1^3 \frac{1}{2} = \frac{1}{3}(3^3 - 1^3)\frac{1}{2} = 4\frac{1}{3} \tag{3.42}$$

as before.

The method of substitution asks us first to identify some new function y, usually a simpler one than the original function. Except in easy cases like this one, it is not always obvious how we should choose this new function. After we have identified the new function, we need to solve this function for the original variable x. Then we differentiate x with respect to y to find the term dx. Finally, we must change the limits of integration, and then we can substitute for the new function, for dx, and for the new limits of integration. We hope that this new integral involving the variable y is easier to perform than the original integral involving the variable x. It is hard to give general rules for how to choose the new variable, and integration by substitution often ends up being done through trial and error.

List of Integrals

$$\int dx = x + c$$

$$\int af(x)dx = a \int f(x)dx$$

$$\int x^n\,dx = \frac{1}{n+1}\,x^{n+1} + c$$

$$\int \sin x \, dx = -\cos x + c$$

$$\int \cos x \, dx = \sin x + c$$

$$\int e^x \, dx = e^x + c$$

$$\int \frac{1}{x} \, dx = \ln x + c$$

$$\int [g(x) + h(x)]dx = \int g(x)dx + \int h(x)dx$$

4. APPLICATIONS

Maximum and Minimum

Largest Area. This situation may not befall any of us, but suppose we had 100 yards of chain link fence. With this fence we want to enclose a rectangle. How long should the sides of this rectangle be if the fenced-in area is to be as large as possible? It could be as small as zero, if we strung out 50 yards and returned the remaining 50 yards next to the first. The question now is what should the dimensions of the sides be so that the rectangle is the largest.

Figure 4.1 shows a rectangle that is 100 yards all around, because if we add up the four sides we do get $x + (50 - x) + x + (50 - x) = 100$. The area

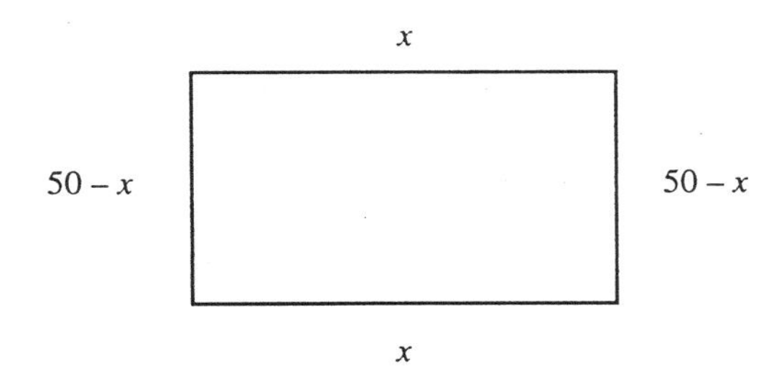

Figure 4.1. Rectangle From 100 Yards of Fence

of this rectangle equals the width multiplied by the height, or $x(50 - x) = 50x - x^2$. We now have the area as a function of the length x of the rectangle. We can write this function as

$$a(x) = 50x - x^2 \tag{4.1}$$

where we use the letter a to stand for area. When $x = 0$ we find $a(0) = 0$, and when we use $x = 50$ we also get $a(50) = (50)(50) - 50^2 = 0$. As might appear obvious, it therefore seems as if we want the side x to be larger than 0 and less than 50.

According to the rules of calculus we can find the minimum or the maximum of $a(x)$ if we differentiate $a(x)$ with respect to x and set the resulting first derivative equal to 0. That gives us an equation that we can solve for x. Sometimes it is not clear whether we get a minimum or a maximum (or an inflection point) of the function for that value of x. For reasons to which we will return, in this case we do get a maximum.

By differentiating the function $a(x)$, setting the resulting function equal to 0, and solving for the unknown x we get

$$a'(x) = 50 - 2x = 0$$

$$x = 25 \tag{4.2}$$

This says that we should make two opposing sides equal to $x = 25$ yards and thus the other two opposing sides equal to $50 - x = 25$ yards. Therefore, because all four sides are equal, the area we want is a square with sides equal to 25 yards. This square encloses the largest area we can get with this fence. For this square the area becomes $a(25) = (50)(25) - 25^2 = 625$ square yards. There is no way we can create a rectangle of more than 625 square yards, no matter how hard we try to lay out the fence.

One reason we know that there is a maximum and not a minimum in this case is that the function equals 0 for the two values 0 and 50 of x, and for any value between 0 and 50 the function is positive. That means the graph of the function goes up from 0 on the left and comes down to 0 again on the right, and an extreme value occurs for an x-value somewhere in between. We have found that this occurs at $x = 25$, and this value has to represent a maximum. Another way to look at this is to notice that as x increases from 0, the derivative is positive for a while. This means the curve goes up. When we get beyond 25, the derivative is negative, and this means the curve goes down. Thus, we must have a maximum and not a minimum at 25. We can also see that if we

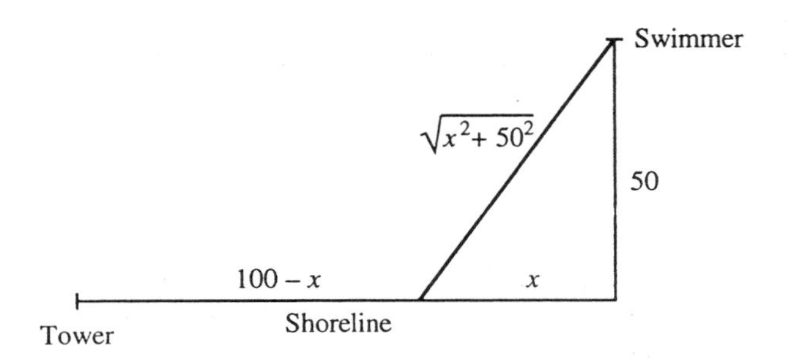

Figure 4.2. Swimmer in Trouble

take the second derivative, we get $a''(x) = -2$. Because the second derivative is negative, we know we have a maximum.

Shortest Time. A lifeguard sits in her tower and watches a swimmer in trouble 50 yards offshore and 100 yards down from the tower. She grabs a little boat and runs down along the shore at 3 yards per second. She then throws the boat in the water and paddles out to the swimmer at 2 yards per second. How far down the shore should she run before getting into the water and starting to paddle if she wants to get to the swimmer in the least possible amount of time?

Figure 4.2 shows the tower, the shoreline, and the troubled swimmer. Suppose the lifeguard runs $100 - x$ yards down the shore first. That way there are x yards left before she is perpendicular to the swimmer from the shoreline. From the Pythagorean theorem we find that she then must paddle $\sqrt{x^2 + 50^2}$ yards diagonally out toward the swimmer. How large should the distance x be for her to get to the swimmer in the shortest time?

Time equals distance divided by speed, so the total time can be written as the function $t(x)$ of x,

$$t(x) = \text{time for running} + \text{time for paddling}$$

$$= \frac{\text{distance for running}}{\text{speed for running}} + \frac{\text{distance for paddling}}{\text{speed for paddling}}$$

$$= \frac{100 - x}{3} + \frac{\sqrt{x^2 + 50^2}}{2} \qquad (4.3)$$

To find the value of x that minimizes the function $t(x)$ we need to differentiate $t(x)$ with respect to x and set the derivative equal to 0. That gives

$$t'(x) = \frac{-1}{3} + \frac{2x}{4\sqrt{x^2 + 50^2}} = 0 \qquad (4.4)$$

Solving this equation for x, we find $x = 44.7$ yards. Thus, she should run $100 - 44.7 = 55.3$ yards before getting into the water. She will end up paddling the boat 67.1 yards.

The time it will take her to reach the swimmer becomes $55.3/3 + 67.1/2 =$ 52.0 seconds. It may be surprising that she runs a shorter distance than she paddles, even though she runs faster than she can paddle. Had she paddled the whole distance from her tower to the swimmer, it would have taken her 55.9 seconds to reach the swimmer. Had she run all 100 yards down the beach and then paddled out to the swimmer, it would have taken her 58.3 seconds. Both these times are greater than the one we found for first running partway down the beach and then paddling out to the distressed swimmer.

Largest Soup Can. When you are in the canned soup business, you must buy sheets of metal to cut up into pieces to make cans. The side of a can comes from a flat, rectangular piece of metal that is rolled to make a tube. When the metal is rolled, the width of the rectangle becomes equal to the circumference of the can and the height of the rectangle becomes equal to the height of the can. The two ends (lids) of the can come from cutting out two circular pieces from another rectangle. Because you have to pay for the metal, the question becomes how to cut the metal so that the cans you make will hold as much soup as possible for a given cost.

Let the can have height h and radius r, as shown in Figure 4.3. Both height and radius are unknown, and with two unknowns we may not be able to solve equations and get exact numerical values for both. This is because we specify only one restriction: that the volume should be as large as possible. Perhaps, however, we can express one unknown in terms of the other and thereby get some sense of what the can should look like.

Because the radius of the top of the can equals r, the diameter of the can becomes $2r$. The circumference of the tube is then $2\pi r$, and with a height of h for the can the area of the flat piece of metal used for the side of the can becomes the width multiplied by the height, or $(2\pi r)(h)$. Similarly, the two round lids for the top and the bottom, each with a diameter of $2r$ and

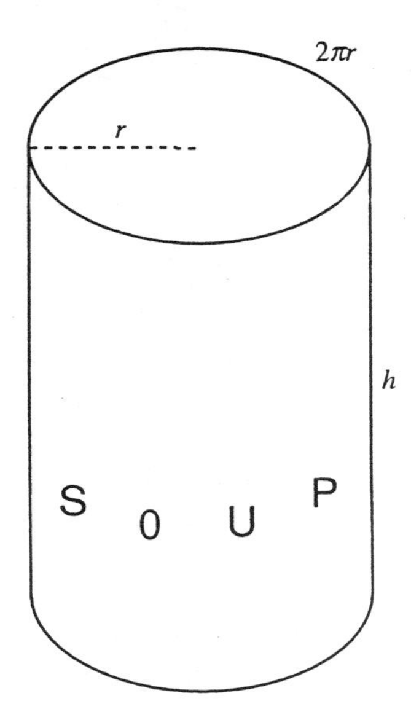

Figure 4.3. A Can With Radius r, Circumference $2\pi r$, and Height h

lying side by side, are cut out of a rectangle that is $4r$ by $2r$. Thus, the total area of the necessary metal becomes

$$Area = (2\pi r)(h) + (4r)(2r) = 2\pi rh + 8r^2 \tag{4.5}$$

With these dimensions the volume of the can is the base of the can times the height. The area of the base is the area enclosed by a circle with radius r, or πr^2, and with the height of h, the volume becomes

$$Volume = \pi r^2 h \tag{4.6}$$

Right now we have both the radius r and the height h as unknowns. Suppose we first look at the volume of the can as fixed. That is, we decide now how much soup we want the can to contain. That means we can solve for the height h, and we find that the height can be expressed as

$$Height = h = \frac{Volume}{\pi r^2} \tag{4.7}$$

As the next step, we substitute this expression for h into the expression for the area. That gives

$$Area = 2\pi r \frac{Volume}{\pi r^2} + 8r^2$$

$$= 2\frac{Volume}{r} + 8r^2 \qquad (4.8)$$

because we can cancel the π and one r in the first term.

Finally, now we have found the area of the metal as a function of the radius of the lid of the can and the fixed volume. We want to use as little metal as possible, so we would like to find the value of r that gives us the smallest area. That implies that we should differentiate this function with respect to r and set the resulting expression equal to 0. That gives

$$\frac{d}{dr} Area = -2\frac{Volume}{r^2} + 16r = 0 \qquad (4.9)$$

By solving for the volume this equation gives us

$$Volume = 8r^3 \qquad (4.10)$$

When we substitute this expression for the volume back into the expression above for the height h (Equation 4.7), we get

$$Height = \frac{Volume}{\pi r^2} = \frac{8r^3}{\pi r^2} = \frac{8r}{\pi} \qquad (4.11)$$

We could stop here, but let us introduce the diameter of the can and work with the diameter instead of the radius. This is because it is easier to measure the diameter of a can than it is to measure the radius. Because the diameter of the lid is twice the radius, or $2r$, the numerator of $8r$ equals 4 times the diameter. We finally get that the relationship between the height and the diameter of our can becomes

$$Height = \frac{4}{\pi} Diameter \approx 1.27\ Diameter \qquad (4.12)$$

Through a further examination of the derivative of the area we can see that we have found a minimum for the area in this case and not a maximum. One way to do that is to take the second derivative with respect to the radius and see that the second derivative is positive for the expression we found for the volume.

Thus, the height of the can should be approximately 27% more than the diameter of the can. An examination of the contents of a kitchen cabinet showed that a typical Campbell's soup can is taller than it should be, a coffee can was just a little bit taller than it should be, and a can of Progresso black beans was right on the mark. Companies that make cat food clearly spend more money on metal for their cans than they need to, because those cans are often wider than they are tall.

Integration

Growth. If we have D dollars in the bank and get p percent compound interest a year, then after one year we will have $D(1+p)$ dollars. Compound interest means that the second year we get interest on the original amount as well as on the interest we accumulated during the first year. After 2 years we will then have $D(1 + p)(1 + p) = D(1 + p)^2$ dollars, and so on for additional years. Suppose, instead, that the bank gave us interest every month instead of every year. We would not get p percent every month; the bank is not that generous. It may give us $p/12$ percent every month, so that after receiving this interest 12 times we would then have

$$D\left(1+\frac{p}{12}\right)^{12} \tag{4.13}$$

dollars after a year.

To carry this argument further, suppose the bank gave us interest n times a year, so that the interest is p/n percent for every interval of time. What happens now as n gets larger? That is, instead of giving us interest every month, we ask what will happen if we get interest every day ($n = 365$), every hour ($n = 365 \times 24 = 8{,}760$), or even every second. Interest every second would mean that we get an interest rate of $p/31{,}536{,}000$ percent paid 31,536,000 times a year.

This means we want to consider the limit of $(1 + p/n)^n$ as n gets very large. What happens here is that in the limit we go from compound interest at *discrete* points in time to *continuous* growth over time.

It can be shown with some mathematical difficulty that with continuous growth the growth factor becomes

$$\lim\left(1+\frac{p}{n}\right)^n = e^p \text{ as } n \text{ goes to infinity} \tag{4.14}$$

where the number $e = 2.718\ldots$ This number is also used as the base for natural logarithms. Thus, D dollars compounded continually for 1 year with p percent interest yields De^p dollars after a year. This type of continuous growth is commonly known as *exponential* growth, because the percentage p appears as an exponent. After t years we would have accumulated $D(t) = De^{pt}$ dollars.

The difference between getting 5% at the end of a year and a continuous growth of 5% is not large after only 1 year. If we start with \$1,000 we will have \$1,050 in the first case and \$1,051.27 in the second case. If we go on with this process, however, then the continuous, exponential growth will be much quicker than the discrete growth.

What happens is that the money grows proportionally to how much money we have at a given time. A small dollar amount D grows slowly, whereas a large amount grows quickly; thus, the rich get richer all the time. Growth proportional to the amount $D(t)$ at a given time can be modeled mathematically by the equation

$$\frac{dD(t)}{dt} = pD(t) \tag{4.15}$$

On the left side we have the growth, as measured by the change in dollars $dD(t)$ per change in time dt. The right side has the constant proportionality factor p times the amount $D(t)$ at time t. Thus, the change is proportional to the amount at any given time. It is not surprising that such an equation is known as a differential equation, because on the left side the function $D(t)$ is differentiated with respect to t. The answer to such an equation is a function $D(t)$ such that the derivative of that function (the left side of the equation) equals the constant p times the function itself (the right side of the equation).

Without going into the finer mathematical details, we find such a function if we rewrite the equation above as the new equation

$$\frac{dD(t)}{D(t)} = p\,dt \tag{4.16}$$

That is, we multiply the dt over on the right side of the equation. To solve the new equation it looks as if we should integrate on both sides, because we have a $dD(t)$ on one side and a dt on the other side. By integrating on both sides we get the equation

$$\int \frac{1}{D(t)} dD(t) = p \int dt \tag{4.17}$$

We recognize that the integral on the left side equals the natural logarithm in $D(t)$ plus a constant, and the integral of dt on the right side equals t plus a constant. The proportionality factor p is a constant and goes outside the integral sign. If we combine the two integration constants into one constant, we get

$$\ln D(t) = pt + constant \tag{4.18}$$

We want to find the dollar amount $D(t)$ as a function of the time t, so we can see how our money grows with time, and we are not as interested in the logarithm of the amount. We do notice that the logarithm of the amount is a linear function of time. This means that if we had data on amounts and time, then we might be able to use linear regression to analyze the relationship between the logarithm of the amount and time.

If we raise both sides of the equation above to exponents of e, we finally get the new equation below with D as a function of time t.

$$D(t) = e^{pt + constant}$$

$$D(t) = e^{constant}\, e^{pt} \tag{4.19}$$

Here we still have an equation that involves the unknown constant, and we need to find the value of this constant. To find this value we need more information.

Suppose we had \$1,000 in the beginning at time 0 and \$3,000 after 15 years. How did the money grow during that period? When we substitute the original amount 1,000 into the equation above for time 0, we get the equation

$$1{,}000 = e^{constant}\, e^{p0} \tag{4.20}$$

Because anything to an exponent of 0 equals 1, we get that

$$1{,}000 = e^{constant} \tag{4.21}$$

By substituting this value for e raised to the constant, we get the dollar amount D as the following function of time t:

$$D(t) = 1{,}000\ e^{pt} \tag{4.22}$$

All that is missing now is the growth rate p, and to find its value we again need more information. This time we use the second piece of information above. We know that at time $t = 15$ years we have $D = \$3{,}000$. By using these numbers we can now solve for the interest p. We substitute the known numbers and get

$$3{,}000 = 1{,}000\ e^{p15} \tag{4.23}$$

This is an equation that has only p as an unknown. We can solve this equation for p by first dividing by 1,000 and taking logarithms again on both sides. We get

$$\ln \frac{3{,}000}{1{,}000} = p15 \tag{4.24}$$

The left side reduces to $\ln 3 = 1.0986$, which gives us

$$p = \frac{1.0986}{15} = 0.0732 \tag{4.25}$$

With 7.32% interest the growth can be expressed in the equation

$$D(t) = 1{,}000\ e^{0.0732t} \tag{4.26}$$

This function fully describes the growth of our original \$1,000.00 and tells us how many dollars we have at any given time. Earlier we only knew how much money we started with and how much we had after 15 years. The additional assumption that the growth was proportional to the change enabled us to get the equation that completely describes the growth over time.

Now we can see how our money grows. After 30 years, for example, we would have

$$D(30) = 1{,}000\ e^{0.0732(30)} = 9{,}000 \tag{4.27}$$

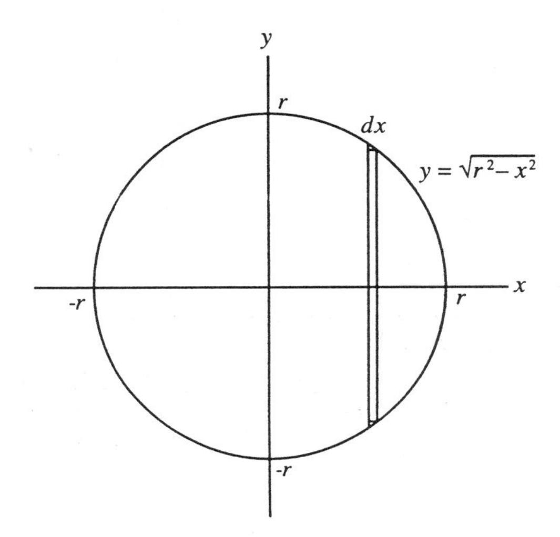

Figure 4.4. A Sphere With Radius *r* and a Slice of the Sphere With Thickness *dx*

This illustrates how exponential growth operates. We started with $1,000.00 and it grew to $3,000.00 after 15 years. In the next 15 years, the amount tripled again, and we now have $9,000.00. If we had received the same interest once a year and compounded, we would have had $8,325.68 instead.

All these results came from the simple assumption that the growth was proportional to the existing amount at any given time. After expressing this fact mathematically, we were able to continue by integrating both sides of an equation. Without integration it would not have been possible to continue beyond that point.

Many problems can be analyzed using exponential growth. Bacteria, for example, often grow exponentially for a while. Similarly, radioactivity decays exponentially, and this enables us to compute how much radioactivity there is at any given time. Many other physical phenomena can be studied this way.

Volume of a Sphere. A tennis ball has a radius r of 1.25 inches. How much volume does it take up? In the more general case, what is the volume of a sphere with radius r? This question can be answered using integral calculus.

Figure 4.4 shows a two-dimensional view of a sphere with radius r, which means we see this sphere as a circle. For any point on the circle with coordinates (x, y) we have from the Pythagorean theorem that $x^2 + y^2 = r^2$. This means that for a fixed value of x the corresponding two values of y on the circle equal $\pm\sqrt{r^2 - x^2}$.

Next we replace the sphere by a series of discs, like coins, of varying sizes from very small on the left and right sides up to one with radius r in the middle. The thickness of each disc equals dx.

The figure shows one such disc. Because the radius of this particular disc equals $\sqrt{r^2 - x^2}$, the surface area of one side becomes π times the square of the radius, or $\pi(r^2 - x^2)$. Because the thickness of the disc equals dx, we get that the volume of this particular disc becomes $\pi(r^2 - x^2)dx$.

The volume of the entire sphere is now the sum of the volumes of these discs from $-r$ on the left to r on the right, in the limit as dx goes to 0. Because integration means adding up many small quantities, we get the volume of the sphere from the integral

$$V = \int_{-r}^{r} \pi(r^2 - x^2)dx$$

$$= \pi(r^2x - \frac{1}{3}x^3)\Big|_{-r}^{r} = \pi\{(r^2r - \frac{1}{3}r^3) - [r^2(-r) - \frac{1}{3}(-r)^3]\}$$

$$= \pi(r^3 - \frac{1}{3}r^3 + r^3 - \frac{1}{3}r^3) = \frac{4}{3}\pi r^3 \tag{4.28}$$

This is the familiar formula for the volume of a sphere with radius r.

Because the radius of a tennis ball is 1.25 inches, the volume of the ball becomes $(4/3)\pi(1.25)^3 = 8.2$ cubic inches. We can also use calculus to find the surface of a sphere with radius r, and the answer is $4\pi r^2$. Thus, the surface of a tennis ball equals 19.6 square inches, which is the same area as a square with sides about 4.4 inches, if we could find some way to open up and flatten the ball.

Distance, Speed, and Acceleration

Differentiation. Moving objects often start from a position standing still. They accelerate for a while until they reach a certain speed and travel a certain distance before they may slow down again and come to a rest.

Calculus can be used to study these three variables, *distance, speed,* and *acceleration,* and the relationships among them.

Often it makes sense to distinguish between how fast an object is moving forward and how fast it is moving backward. We can make this distinction by replacing the term *speed* by the term *velocity,* because we permit velocity to be either positive or negative depending on the direction of travel, while speed is typically measured by a positive number no matter whether the object travels forward or backward.

Suppose we drive a car at a steady 60 miles per hour on the expressway. After driving for t hours, the distance covered by the car becomes $60t$. Thus, the distance is a function of the time, and this can be expressed in the equation

$$Distance(t) = 60t \tag{4.29}$$

The graph of this equation is a straight line starting at the origin and pointing up and to the right with a slope of 60.

If we differentiate the distance function, we get the slope of the tangent to the curve at any point. Because the vertical axis is measured in miles and the horizontal axis is measured in hours, the slope as measured by the "rise" over the "run" will be measured by miles/hour. This is typically abbreviated by *mph* and gives the speed of the car (we use the term *speed* here because the car is only moving in one direction). Thus, we get the speed by differentiating the distance function with respect to time. Here we get

$$Speed(t) = Distance'(t) = 60 \tag{4.30}$$

In this case the speed is a constant and therefore not a function of time, but we still can think of the speed as a function of time even though here the function is simply a constant. With more complicated distance functions we often get the speed as a function that involves t. Here, because we already knew that the car was traveling 60 mph, we have not learned anything new so far. We do see how we find the speed of a moving object by differentiating the distance function.

Next, let us differentiate the speed function with respect to t. This will be the same as taking the second derivative of the distance function. Here we get

$$Speed'(t) = Distance''(t) = 0 \quad (4.31)$$

The unit of the speed variable is miles/hour, and the unit of the time variable is hours. When we differentiate the speed function we get the slope of the tangent to the curve. The "rise" over the "run" for this slope becomes (miles/hour)/hour, and this tells us how the speed changes with time. Such a speed change is also known as the acceleration of a moving object. Here, because the car is moving at a constant 60 miles/hour, there is no change in the speed and the acceleration equals zero. We note that the unit for acceleration can also be written miles/(hour)^2.

Thus, the three quantities *acceleration, speed,* and *distance* for a moving object are related according to the following expression:

$$Acceleration(t) = Speed'(t) = Distance''(t) \quad (4.32)$$

We go from *distance* to *speed* to *acceleration* by successive differentiations. We can also go the other way from *acceleration* to *speed* to *distance* by successive integrations.

Integration. It follows from the previous section on differentiation that we can go from one equation to another by integration. Specifically, we get the speed equation by integrating the acceleration equation, and we get the distance equation by integrating the speed equation. That is,

$$Speed(t) = \int Acceleration(t)dt$$

$$Distance(t) = \int Speed(t)dt \quad (4.33)$$

As an example, let us look at a reasonably fast car that takes 8 seconds to reach 60 miles per hour under maximum acceleration. One thing we want to know is how far the car needs to travel to reach that speed.

Let us do the computations in feet and seconds, such that 60 miles per hour becomes 88 feet per second. We also assume that the car accelerates with a constant but unknown acceleration of a ft/sec^2. This means that for acceleration we have the equation

$$Acceleration(t) = a \quad (4.34)$$

To find the speed at any given time t we need to integrate the acceleration equation. That gives us the speed as a function of the time t as

$$Speed(t) = \int a\, dt = at + constant \tag{4.35}$$

So far the constant is unknown. To find the value of the constant, we know that at time $t = 0$ the speed of the car is 0, because at that time it is standing still. Thus, for that point in time we have

$$0 = a\,(0) + constant \tag{4.36}$$

and this means the constant equals 0. Thus, for the speed we now have

$$Speed(t) = at \tag{4.37}$$

To find the unknown acceleration a we now make use of the fact that we know the speed to be 88 ft/sec at time t = 8 seconds. Substituting these numbers into the equation above gives us an equation we can solve for the unknown acceleration a. We get

$$88 = a\,8$$

$$a = 11 \text{ ft/sec}^2 \tag{4.38}$$

That gives us the speed equation $Speed(t) = 11t$. We want the distance traveled, so we have to integrate the speed equation to get the distance equation. By integrating we get

$$Distance(t) = \int 11t\, dt = \frac{1}{2}\, 11\, t^2 + constant \tag{4.39}$$

This constant is also unknown, but because at time $t = 0$ the distance is also equal to 0, the constant has to be equal to zero for the equation to be satisfied.

This finally leaves us with the distance equation we wanted:

$$Distance(t) = 5.5t^2 \tag{4.40}$$

In particular, we want to know the distance at $t = 8$ seconds, and now we get

$$Distance(8) = 5.5\ (8^2)\ \text{ft} = 352\ \text{ft} \tag{4.41}$$

This distance is almost 120 yards, so the car needs a distance of somewhat more than the length of a football field to reach a speed of 60 miles per hour. This little computation may be useful the next time we want to enter an expressway where the cars travel at 60 miles per hour and we want to merge in with the traffic at that speed. If the entry ramp is not long enough, we will not be able to reach the speed we want. Most of us never realized that highway engineers use calculus to design expressway entry ramps.

In this example we see that the speed increases linearly with time, so in 10 seconds we will have reached a speed of 75 miles per hour. It will take us a distance of 550 feet, or an additional 200 feet, to reach that speed.

Gravity. An object dropped from, say the Leaning Tower of Pisa, will be pulled toward the ground with increasing speed because of the effect of Earth's gravitational force. For the same reason, an object shot up into the air will gradually lose speed until it comes to a complete stop, and then it will fall down again with increasing speed. If we disregard air resistance for a moment, the gravitational force of Earth is such that acceleration and deceleration equal about 32 ft/sec^2 for any object, be it a feather or a bullet. This number will be positive for movement in one direction and negative for movement in the opposite direction.

Suppose an object is propelled straight up at an initial velocity of 1,000 feet per second. In this case we prefer the term *velocity* instead of *speed* because the object will first move upward and then it will change direction and move downward again. Let us study how the distance traveled depends on the length of time the object is in the air. Starting with the acceleration a, we have the acceleration as a function of time according to the equation

$$a(t) = -32 \tag{4.42}$$

We use a negative value for the acceleration because the object will be losing speed as it moves upward.

To find the velocity v we need to integrate the equation for the acceleration. We get

$$v(t) = \int -32dt = -32t + constant \tag{4.43}$$

To find the value of this constant we make use of the fact that the initial velocity was 1,000 feet per second. This is the velocity at the time $t = 0$, which means the constant has to equal 1,000. Thus,

$$v(t) = -32t + 1{,}000 \tag{4.44}$$

We see from this equation that as time increases, the velocity will decrease. After 10 seconds, for example, the original velocity of 1,000 feet per second will have been reduced to –320 + 1,000 = 680 feet per second.

From this equation for the velocity we can find the distance function by one more integration. We get

$$d(t) = \int(-32t + 1{,}000)dt = -16t^2 + 1{,}000t + constant \tag{4.45}$$

At time 0 the distance traveled equals 0, which means that this constant equals 0. Finally we get for the distance that

$$d(t) = -16t^2 + 1{,}000t \tag{4.46}$$

How long does it take for the object to reach its maximum height before it starts to fall down again, and how far will it have traveled at that point? The object reaches its maximum height at the fleeting moment when the velocity finally equals 0, before it starts falling down again. Equation 4.44 describes the velocity. We set velocity equal to 0 and get

$$-32t + 1{,}000 = 0$$

$$t = 31.25 \text{ sec} \tag{4.47}$$

From Equation 4.46, we see that after $t = 31.25$ seconds the object will have reached a height of

$$d(31.25) = -16(31.25)^2 + 1{,}000(31.25) = 15{,}625 \text{ ft} \tag{4.48}$$

or about 3 miles.

Applications in Statistics

Expected Value and Empirical Mean. When we have a sample of observations from a continuous random variable, we can form the empirical histogram to get a sense of the distribution of our data. Commonly, we assume that the sample comes from an infinitely large population. For such a population the histogram can be shown as a smooth curve if we make the intervals short enough, and we say that the variable follows the distribution shown by this curve. The equation for such a curve can be denoted $f(x)$, and the curve is known in statistical theory as the density function of the variable. Often this distribution is assumed to be a normal distribution.

We find the mean $\bar{x}$ of the sample data by adding up all the observations and dividing by the number of observations. This does not work for the population if it has infinitely many observations. Instead, the mean value μ of the variable for the population is called the expected value of the variable. The expected value of a variable is defined as the integral

$$\mu = \int_{-\infty}^{\infty} x f(x) dx \tag{4.49}$$

This integral tells us to multiply each value x of the variable by the height of the density function $f(x)$ at that value times dx and then add all these products. How does the computation of μ compare with the computation of $\bar{x}$?

We find the sample mean from the formula

$$\bar{x} = \frac{x_1 + x_2 + x_3 + \ldots + x_n}{n} \tag{4.50}$$

Suppose now that not all the x values are different, but that we have several observations with the same value. If we have n_1 values of x_1, we can either add up x_1 n_1 times or we can find this sum by multiplying x_1 by n_1. That way, the sample mean can be written

$$\bar{x} = \frac{n_1 x_1 + n_2 x_2 + n_3 x_3 + \ldots}{n}$$

$$= \frac{n_1}{n} x_1 + \frac{n_2}{n} x_2 + \frac{n_3}{n} x_3 + \ldots$$

$$= p_1x_1 + p_2x_2 + p_3x_3 + \ldots = \sum_{i=1}^{n} p_ix_i \qquad (4.51)$$

where $p_i = n_i/n$ is the proportion of the observations that has the value x_i. This way the sample mean is found as the sum of the products of each observed value of x times the proportion of times this value occurs in the sample.

With infinitely many values, a proportion of observations is known as a probability. The probability of getting an observation that lies in a small interval between the value x and the value $x + dx$ equals the area under the curve of $f(x)$ from x to $x + dx$. This area can be approximated by a rectangle with base dx and height $f(x)$, and the area of such a rectangle becomes $f(x)dx$. Thus, the probability $f(x)dx$ takes the place of the proportion p_i in the expression for $\bar{x}$.

We find the sample mean by adding the products of the proportion of observations in the sample with a particular value and that value of the variable, or the sum of the terms p_ix_i. Similarly, we now find the population mean by adding the products of the probability of a value in a small interval times a value in the interval, or the sum of terms of the form $[f(x)dx]x$. This is like the sum of the terms of the form p_ix_i above, but when the variable is continuous and we have infinitely many values of the variable, such a sum is by definition written as an integral.

Thus, the population mean μ is found from the formula

$$\mu = \int_{-\infty}^{\infty} xf(x)dx \qquad (4.52)$$

and we see how this formula has a direct counterpart in the formula written above for a sample mean as a sum of products of proportions and x values.

For example, for the normal distribution we get the mean by finding the integral

$$\int_{-\infty}^{\infty} x\frac{1}{\sqrt{2\pi}\sigma} e^{\frac{-(x-\mu)^2}{2\sigma^2}} dx = \mu \qquad (4.53)$$

This integral is not easy to find, and we do not go through that integration here.

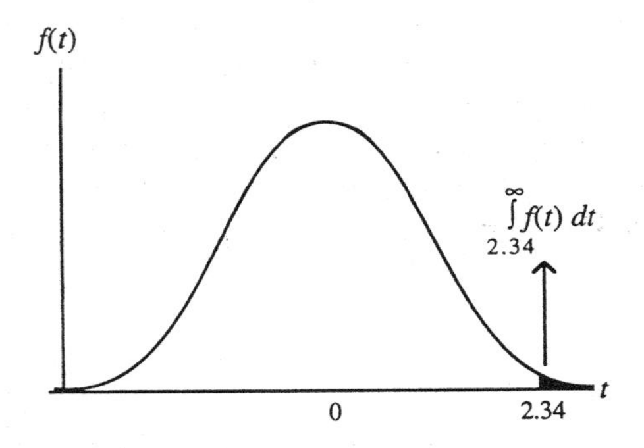

Figure 4.5. Statistical p Value as an Integral

Similarly, the population variance σ^2 is found from the formula

$$\sigma^2 = \int_{-\infty}^{\infty} (x - \mu)^2 f(x) dx \tag{4.54}$$

This formula also has a direct counterpart in the formula we use to define the sample variance s^2.

Statistical p value. Another place in statistics where we often make use of calculus is in the computation of a statistical p value in hypothesis testing. Any time we use a normal, t, chi-square, or F distribution to find a p value, we use calculus.

Suppose we have a null hypothesis that says a parameter equals a certain value. From the observed data we find $t = 2.34$ on 21 degrees of freedom as the value of the test statistic for the null hypothesis. The p value tells how often we would get the sample resulting in our value of t or samples resulting in more extreme values of t than the one we have, drawing from a population for which the null hypothesis is true. This is the same as finding how often we would get a value of t equal to or larger than 2.34 if we sampled many values of t from the t distribution with 21 degrees of freedom.

Figure 4.5 shows the typical graph we see in a statistical table. The curve shows the distribution of t with 21 degrees of freedom, and we want the probability that $t > 2.34$. The total area under the curve equals 1.00, and the shaded area on the right represents the probability we want.

According to the rules of integration, we find this area by integrating the function for the curve from 2.34 to infinity. This can be expressed mathematically as

$$p\ value = \int_{2.34}^{\infty} f(t)dt \tag{4.55}$$

When we substitute the proper function for $f(t)$ we get

$$p\ value = \int_{2.34}^{\infty} 0.39\left(1 + \frac{t^2}{21}\right)^{-11} dt = 0.015 \tag{4.56}$$

The numerical result tells us that only 15 out of 1,000 times would we get a value of t larger than 2.34. Any time we find a statistical p value from one of the continuous statistical distributions, we use calculus to perform an integration.

This particular integral is not easy to evaluate. We need to find a function $F(t)$ that has the $f(t)$ above as its derivative, but this is mathematically difficult. The variable of integration is t, but in the function we have a t^2 term that, because of the negative exponent, occurs in a denominator where we have $1 + t^2/21$. Because of the difficulty of performing this integration, we have statistical tables for the t distribution or use computer software to find the answer instead. What we can do to find the area under the curve is to go back to basics and divide the area up into little rectangles and add those rectangles. Such numerical methods have to be used with a computer when there is no direct solution for the integral.

Total Area Under a Curve. We showed above that the t distribution with 21 degrees of freedom can be expressed in the formula

$$f(t) = 0.39\left(1 + \frac{t^2}{21}\right)^{-11} \tag{4.57}$$

This is the equation for a bell-shaped curve that is symmetric around 0 and stretches out to plus and minus infinity. Similar formulas exist for the t variable with other degrees of freedom as well as for the normal, the chi-square, the F, and other statistical distributions.

One thing all these formulas have in common is that any area under the curve represents the probability that the variable falls in the corresponding interval on the horizontal axis. Because the total probability for any variable equals 1, we now know that this fact can be expressed using calculus. A total probability of 1 is equivalent to saying that the integral of the curve over all values of the variable equals 1, such that the total area under the curve equals 1. Mathematically this can be expressed

$$\int_{-\infty}^{\infty} 0.39\left(1+\frac{t^2}{21}\right)^{-11} dt = 1 \tag{4.58}$$

The important part of this t distribution is the expression inside the parentheses. The only reason we have the constant 0.39 is to ensure that the integral integrates to 1. Most distributions have a constant like this as part of the formula for the curve to ensure that the function integrates to 1. The standard normal distribution, for example, has the factor $1/\sqrt{2\pi} = 0.40$ in the front to make the integral of the distribution from minus infinity to infinity equal to 1.

Mean or Median. The choice of whether to use the mean or the median of a set of observations is not always easy to make. Suppose our data consist of n observations $x_1, x_2, x_3, \ldots, x_n$. Should we summarize these data by the mean or the median? Calculus provides some information about the difference between the two.

The purpose of any summary measure in statistics is to pick a constant number c and see how well we can replace all the original observations by this new number. The distance from an observation x_i to c becomes the difference $x_i - c$. We can find this difference for each observation. If all these differences are small, then c is close to all the observations and c would be a good representation of our data. The only remaining question is how we should pick c.

One possibility is to *square* all the differences and then add all the squares. That way we get the sum

$$(x_1 - c)^2 + (x_2 - c)^2 + (x_3 - c)^2 + \ldots + (x_n - c)^2 \tag{4.59}$$

The only thing we do not know about this sum is the value of c. All the x values come from the data, and c is the only unknown. That means this sum is a function of c. If we could find the value of c that would make this sum the smallest, then we know that each square would be small, because all the squares are positive, and therefore each difference would be small. Thereby, c would in some sense be in the middle of all the data and close to all the observations. At least, no other constant value would be closer to the observations than c, even though the sum of squares around c may not be small in an absolute sense.

To find the value of c that makes this sum a minimum, we need to take the derivative of the sum with respect to c and set the derivative equal to 0. That gives us

$$2(x_1 - c)(-1) + 2(x_2 - c)(-1) + 2(x_3 - c)(-1) + \ldots + 2(x_n - c)(-1) = 0 \qquad 4.60$$

We differentiated each term using the chain rule. First we took the derivative of the outside function, which gives $2(x_i - c)$. Then we multiply this by the derivative of the expression inside the parentheses with respect to c, which gives us -1.

To solve the equation above for c we can first cancel out all the 2s and -1s. When we open all the parentheses we get

$$(x_1 + x_2 + x_3 + \ldots + x_n) - (c + c + c + \ldots + c) = 0$$

$$nc = x_1 + x_2 + x_3 + \ldots + x_n$$

$$c = \frac{x_1 + x_2 + x_3 + \ldots + x_n}{n} = \bar{x} \qquad (4.61)$$

This means that if we want to summarize the data by the constant value of the variable that gives us the smallest sum of squared differences between the observations and that constant, then we should use the mean.

Another way to find a constant close to all the observations is to require that the *absolute values* of the differences between the observations and the constant should be small. One way to achieve this goal is to find a constant c such that the sum of all these absolute values is small. This is because if we know that the sum is small, then each of the terms is small because all the absolute values are positive. Thus, we want to consider the sum

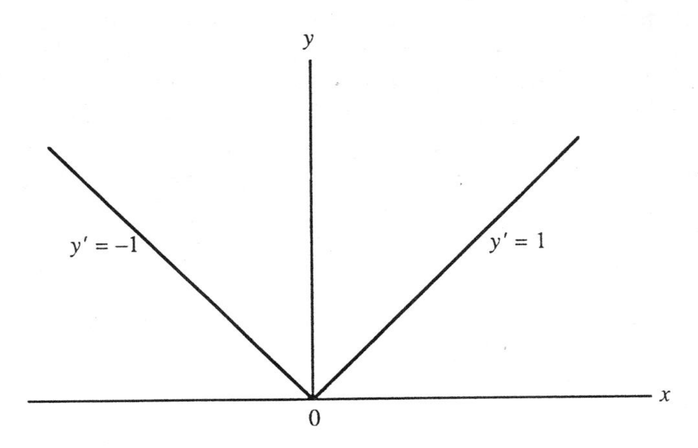

Figure 4.6. Graph of the Absolute Value Function $y = |x|$

$$|x_1 - c| + |x_2 - c| + |x_3 - c| + \ldots + |x_n - c| \tag{4.62}$$

Because all the data are known, this sum depends only on c. We want to find the value of c that makes this sum as small as possible. To do that we can take the derivative of the sum with respect to c and find the value of c that sets this derivative equal to 0. Figure 4.6 shows a graph of the absolute value function, together with the derivative of the absolute value function. From the graph we see that the derivative equals +1 if the term is positive and −1 if the term is negative. At 0 the function makes a 90 degree turn, and at that point there exists no derivative.

Thus, the derivative of each term in the sum of absolute values is either +1 or −1, depending upon whether the term is positive or negative. Thus, we can write the derivative of the entire sum as

$$\pm 1 \pm 1 \pm 1 \pm \ldots \pm 1 \tag{4.63}$$

This sum will equal 0 if the number of +1s equals the number of −1s. With an even number of observations we can achieve this by choosing c such that half of the differences $x_i - c$ are positive and the other half are negative. Each positive difference has a derivative of +1, and each negative difference has a derivative of −1. Thus, c should be chosen such that it is larger than one half of the observations and smaller than the other half. Such a quantity is known as the median value of the variable. With an odd number of observations c is also equal to the median of the data, which in that case equals the middle observation. All we do then mathematically is exclude the observation that is equal to c, because the difference for that observation equals zero and we have no derivative of the absolute value function at zero.

This implies that c becomes the median of our data, because the median divides the data into two equal parts such that half of the observations are smaller than the median and half of the observations are larger than the median. Thus, we should use the median when we want a value of the variable that minimizes the sum of the absolute values of the differences between the observations and the constant. This implies we should use the median instead of the mean when we do not want to give a large influence to extreme observations, either large or small. The reason for this is that an observation that is far away from the constant c will contribute more to the sum of squared differences than to the sum of absolute differences, because large numbers when squared become even much larger.

Another way to look at the difference between the mean and the median is in terms of the following game. Suppose we are to guess what a particular value in our data is equal to. One strategy would be to guess that the value is equal to the mean. We know that this guess is likely to be wrong, but we hope that we will not be wrong by very much because the mean lies somewhere in the middle of the data. Suppose the penalty we are assessed for our mistake is the square of the error. That is, the correct value is x_i and we guess $\bar{x}$, and so the penalty becomes $(x_i - \bar{x})^2$. If we play this game for each observation, then our total penalty, or loss, from this game becomes the sum of these squares. Could we have played this game better by choosing some number other than the mean? The answer is no, because we know it is the mean that minimizes the sum of squares of the differences around a particular value.

Let us change the game such that the loss we suffer from a wrong prediction is the absolute value of the difference between the true value and the value we guess, instead of the squared difference. If we go through all n observations, such that the total loss is the sum of these absolute values, then we know that our best strategy is to predict the median of the observations each time. This is because it is the median that minimizes the sum of the absolute values of the differences around a particular value.

Thus, by the use of calculus, the choice between the median and the mean comes down to whether our loss function is the sum of the squares of the mistakes or the sum of the absolute values of the mistakes we make in predicting the data from the summary value.

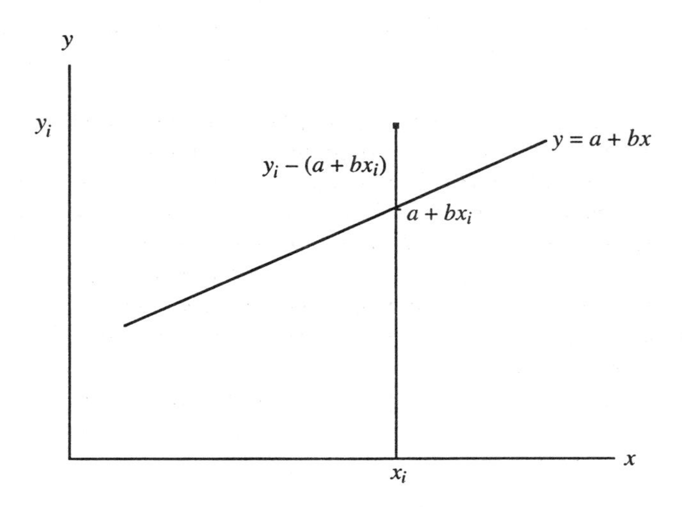

Figure 4.7. Regression With an Observed Point and a Residual

Regression Analysis. Regression analysis is another example of a statistical method that relies heavily on calculus. Figure 4.7 shows x and y as the two coordinate axes, and in this scatterplot there is one point with coordinates (x_i, y_i) and the regression line with equation $y = a + bx$.

To find the regression line we first consider the vertical distance from an observed point to the regression line. This is the distance marked on the figure as $y_i - (a + bx_i) = y_i - a - bx_i$. The regression line is the line that has the smallest sum of squared distances from the points to the line. That is, we find the vertical distance from each point to the line, square all these distances, and then add all the squares. No other line will give a smaller (residual) sum than the regression line.

This least squares principle is used to find the line, or to find the formulas for the computation of the intercept a and the slope b. Here is how these formulas are found. We want to minimize the residual sum of squares

$$(y_1 - a - bx_1)^2 + (y_2 - a - bx_2)^2 + (y_3 - a - bx_3)^2 + \ldots + (y_n - a - bx_n)^2 \quad (4.64)$$

The xs and ys are the known data values, so the sum depends only on the two unknowns a and b. Any time we substitute values of a and b we get a value for this sum of squares. The trick now is to find the particular a and b that will give the smallest possible sum. Two unknowns are no different from one unknown, so we take the derivative of the sum above first with respect to a and then with respect to b. We set these two derivatives equal to zero, because we are looking for a minimum, and then solve the resulting two equations for a and b.

To take the derivative of the sum we must use the chain rule on each square in the sum. First we take the derivative of the square, and then we multiply that by the derivative of what is inside the parentheses. The derivative of the square is 2 times what is inside the parentheses. The derivative of the inside with respect to a is just -1, and the derivative of the inside with respect to b is the constant x_i. Taking the two derivatives and setting them equal to zero gives us the two equations

$$2(y_1 - a - bx_1)(-1) + 2(y_2 - a - bx_2)(-1) + \ldots = 0 \qquad (4.65)$$

$$2(y_1 - a - bx_1)(-x_1) + 2(y_2 - a - bx_2)(-x_2) + \ldots = 0 \qquad (4.66)$$

Next we can cancel all the 2s and all the -1s. Then, after opening the parentheses and rearranging some of the terms, the equations become

$$(a + a + \ldots + a) + b(x_1 + x_2 + \ldots + x_n)$$

$$= y_1 + y_2 + \ldots + y_n \qquad (4.67)$$

$$a(x_1 + x_2 + \ldots + x_n) + b(x_1^2 + x_2^2 + \ldots + x_n^2)$$

$$= x_1y_1 + x_2y_2 + \ldots + x_ny_n \qquad (4.68)$$

The equations can be written in a more compact form if we use the summation sign Σ. That gives us

$$na + b\sum x_i = \sum y_i \qquad (4.69)$$

$$a\sum x_i + b\sum x_i^2 = \sum x_iy_i \qquad (4.70)$$

In the first equation, if we divide both sides by n and solve for a, we get the usual formula for the intercept in terms of the slope and the two means,

$$a = \bar{y} - b\bar{x} \tag{4.71}$$

When we solve the two equations for the other unknown *b*, we get for the slope

$$b = \frac{n\sum x_i y_i - \sum x_i \sum y_i}{n\sum x_i^2 - \left(\sum x_i\right)^2} \tag{4.72}$$

This is the common computing formula for the slope of the regression line. After a few algebraic manipulations the slope can also be written as the common definitional formula,

$$b = \frac{\sum (x_i - \bar{x})(y_i - \bar{y})}{\sum (x_i - \bar{x})^2} \tag{4.73}$$

Simple regression is the most common example of least squares analysis in statistics. Here we have found the *a* and *b* that minimize the residual sum of squares (even though we have not shown that we have a minimum here). The slope and intercept are used to find the regression line, as well as the various sums of squares in regression.

The same principles apply in multiple regression analysis. The only difference is that there are more parameters to work with and more equations to solve. The necessary equations are therefore often written out in matrix form. Analysis of variance is also based on the same principles of least squares, and both fall under what is known as the general linear model.

Differential calculus provides us with a new look at regression coefficients. Let us use a linear regression model with two independent variables as an example. The regression equation can then be written as

$$y = a + b_1 x_1 + b_2 x_2 \tag{4.74}$$

The typical interpretation of the partial regression coefficient b_1 is that it tells us how much change there will be in the dependent variable y for a one-unit change in the independent variable x_1 while the second independent variable x_2 is held constant.

From a mathematical point of view we have an equation in which one variable y is a linear function of two other variables x_1 and x_2. Suppose we now differentiate this function with respect to the first variable x_1. Because we could have differentiated with respect to the second variable as well, we need the partial derivatives (denoted by the symbols $\partial y/\partial x_1$ and $\partial y/\partial x_2$).

In the case of the partial derivative of y with respect to x_1, a is a constant and so is the term b_2x_2. Because derivatives of constants equal zero, we get the partial derivative

$$\frac{\partial}{\partial x_1} = b_1 \tag{4.75}$$

and similarly for the partial derivative of y with respect to x_2. Thus, a particular regression coefficient for an independent variable can also be interpreted as the partial derivative of the dependent variable with respect to the corresponding independent variable. As such, the partial regression coefficient tells us the instantaneous change in y for a change in x_i.

Applications in the Social Sciences

The social sciences have long looked at the uses of mathematics in the natural sciences and wished that similar uses of mathematics would be possible for them. That would establish the social sciences as *real* sciences in the eyes of the world, and it would aid in the discovery of new findings in the social sciences the same way mathematics has aided the natural sciences.

Economics is the one social science where this goal has been pursued with substantial success. This is not surprising, given that many economic variables are continuous interval or ratio scale variables and that economists are interested in changes that take place in these variables.

People working in other social sciences have tried to introduce mathematics into various fields, as exemplified by books by Alker, Deutsch, and Stoetzel (1973) on mathematics in political science; Coleman (1964), Fararo (1973), and Hamblin, Jacobsen, and Miller (1973) in sociology; and Coombs (1983) in psychology. This is not a movement that has left many marks in the literature, and mathematical articles are rare in the major social science journals.

The Arms Race. One article that makes use of calculus is by Hamblin, Hout, Miller, and Pitcher (1977). The authors begin with Richardson's famous improved model for the arms race, which they state in these two equations:

$$\frac{dy}{dt} = kx - \alpha y \quad \text{and} \quad \frac{dx}{dt} = ly - \beta x \tag{4.76}$$

The left side of the first equation represents the change in armaments over time by nation Y and, similarly, the left side of the second equation represents the change in armaments over time for nation X; k and l are constants called the insecurity coefficients; x is the cumulative armaments for nation X and y the cumulative armaments for nation Y; and α and β are known as fatigue coefficients.

These equations are differential equations that can be solved through the use of integration for x and y as functions of time t. After showing the solutions, the authors discuss a modification of the model and derive their own solution equations for x and y as functions of time. They then fit their solutions to data from several different arms races and discuss the policy implications of their model.

Coefficients in Dynamic Models. Several examples of the use of calculus are discussed by Nielsen and Rosenfeld (1981). Their main concern is with the way in which the coefficients in the mathematical models are interpreted.

One of the models they discuss is of the process of political mobilization, and it can be written as a differential equation of the form

$$\frac{dy}{dt} = b(y - c)x \tag{4.77}$$

where y denotes electoral support for a party in a district, such that the left side of the equation gives the rate of change of support over time t (rate of mobilization). The variable x represents the social composition of the district, for simplicity here measured by one variable, for example, the proportion of blue-collar workers in the district.

From there it is possible through integration of both sides of the equation to solve the differential equation for y as a function of time and come up with an equation for $y(t)$. That way we get the trajectory of electoral support over time. The parameters in a model like this one are often more meaningful in the expression for the trajectory $y(t)$ than they are in the expression for the instantaneous change dy/dt. The authors point out that in this case the equation for the trajectory makes it possible to decompose the process of political mobilization into two components. One component describes the speed of the process away from the initial level of mobilization, and the other component describes the factors that affect the value of the mobilization reached at a time of equilibrium.

Demography. Demographic variables often measure sizes of populations, rates of events such as births and deaths, and changes in these variables. Because these variables can be thought of as continuous interval or ratio scale variables, like many economic variables, it is not surprising that demographers, like economists, have made use of calculus to advance their field. One such example is by Preston (1982).

Preston starts by pointing out that there is a difference between the distribution for a population and the experience of an individual over a lifetime for a characteristic like marital status. The expected proportions of life spent in different marital statuses for an individual is typically different from the proportions of a population in those statuses, even though the population proportions have been aggregated from individuals. The article deals with these differences.

Let there be some attribute in the population that is differentially distributed by age. First, we assume we have a stable population, meaning that age-specific birthrates and death rates have been stable for a long time. It can then be shown that the proportion of the population of age *a* can be found from the expression

$$c(a) = \frac{e^{-ra}p(a)}{\int e^{-ra}p(a)da} \tag{4.78}$$

where *r* is the growth of the population and *p*(*a*) is the probability of an individual surviving from birth to time *a*. In this expression we recognize the term *e* to the exponent $-r$ as the effect of continuous growth. In addition there is a minus sign here, meaning that there is negative growth, in the sense that a population decreases over time. The numerator gives the number of people at age *a*. The denominator is a sum over *a* of the number of people at age *a*. If time were measured in years, then we would be interested in the number of people of that given age, and the denominator would have been a sum instead of an integral. We would have added from 0 to 100, say, to get the total population, and the numerator would have been the number of people of a given number of years. That way, the fraction above would have been the proportion of people of a given age.

If we think of age as a continuous variable and we want the proportion of people in the population that are of a given age, then we get an integral instead of a sum in the denominator. In addition, we are not particularly interested in what the numerical value of $c(a)$ might be for some a; we are more interested in the expression of $c(a)$ as a function of a.

From his analysis using the expression above and other, similar expressions, Preston is able to show that fertility differences affect populations but not life cycles. Also, mortality differences affect life cycles but not populations.

Conclusion

These examples show a few of the many uses to which calculus has been put. The invention of calculus was motivated by its potential applications. At the same time, mathematicians are mainly concerned with how to differentiate and integrate mathematical functions without necessarily much regard for what kinds of phenomena these functions might represent in the real world.

Other people use calculus to derive new knowledge in their substantive fields. That means they first have to create the necessary mathematical functions as models for the phenomena they study. Then they can go ahead and apply calculus methods to these mathematical models and learn from the results produced by the application of calculus. This initial step of creating the appropriate models is often a difficult one.

The nature of calculus is such that it is well suited for the study of dynamic processes. This is because the ratio dy/dx represents change in one variable produced by another variable. The model for the arms race is an example of such change. There the change in arms with respect to time is assumed to follow a certain form. Many other social phenomena are also expressed as change over time, and this may be an area that will see a greater use of calculus. Ideally, we should have a process that is expressed as a social theory in axiom form, and then these axioms should be such that they can be translated into mathematical models that then can be manipulated by the use of calculus.

REFERENCES

ALKER, H. R., DEUTSCH, K. W., and STOETZEL, A. H. (Eds.). (1973) *Mathematical Approaches to Politics.* San Francisco: Jossey-Bass.

COLEMAN, J. S. (1964) *Introduction to Mathematical Sociology.* New York: Free Press of Glencoe.

COOMBS, C. H. (1983) *Psychology and Mathematics.* Ann Arbor: University of Michigan Press.

FARARO, T. J. (1973) *Mathematical Sociology: An Introduction to Fundamentals.* New York: John Wiley.

HAMBLIN, R. L., JACOBSEN, R. B., and MILLER, J. L. L. (1973) *A Mathematical Theory of Social Change.* New York: Wiley-Interscience.

HAMBLIN, R. L., HOUT, M., MILLER, J. L. L., and PITCHER, B. L. (1977) "Arms races: A test of two models." *American Sociological Review* 42: 338-354.

NIELSEN, F., and ROSENFELD, R. A. (1981) "Substantive interpretations of differential equation models." *American Sociological Review* 46: 159-174.

PARKER, S. (Ed.). (1992) *McGraw-Hill Encyclopedia of Science and Technology* (7th ed., vol. 3). New York: McGraw-Hill.

PRESTON, S. H. (1982) "Relations between individual life cycles and population characteristics." *American Sociological Review* 47: 253-264.

ABOUT THE AUTHOR

GUDMUND R. IVERSEN is Professor of Statistics and Director of the Center for Social and Policy Studies at Swarthmore College. He received M.A. degrees in mathematics and sociology from The University of Michigan and his Ph.D. in statistics from Harvard University. He is interested in statistical education as well as Bayesian statistics, contextual analysis, and the application of statistics to the social sciences in general. He is the author or coauthor of three other volumes in this series: *Analysis of Variance, Bayesian Statistical Inference,* and *Contextual Analysis.* At Swarthmore he is a member of the Department of Mathematics and Statistics, in which he has occasionally taught sections on calculus.

71 **Analyzing Complex Survey Data** *by Eun Sul Lee, Ronald N. Forthofer & Ronald J. Lorimor*
70 **Pooled Time Series Analysis** *by Lois W. Sayrs*
69 **Principal Components Analysis** *by George H. Dunteman*
68 **Rasch Models for Measurement** *by David Andrich*
67 **Analyzing Decision Making** *by Jordan J. Louviere*
66 **Q-Methodology** *by Bruce McKeown & Dan Thomas*
65 **Three-Way Scaling and Clustering** *by Phipps Arabie, J. Douglas Carroll & Wayne S. DeSarbo*
64 **Latent Class Analysis** *by Allan L. McCutcheon*
63 **Survey Questions** *by Jean M. Converse & Stanley Presser*
62 **Information Theory** *by Klaus Krippendorff*
61 **Multiple Comparisons** *by Alan J. Klockars & Gilbert Sax*
60 **Linear Programming** *by Bruce R. Feiring*
59 **Meta-Analysis** *by Fredric M. Wolf*
58 **Randomized Response** *by James Alan Fox & Paul E. Tracy*
57 **Understanding Regression Analysis** *by Larry D. Schroeder, David L. Sjoquist & Paula E. Stephan*
56 **Introduction to Linear Goal Programming** *by James P. Ignizio*
55 **The Logic of Causal Order** *by James A. Davis*
54 **Multivariate Analysis of Variance** *by James H. Bray & Scott E. Maxwell*
53 **Secondary Analysis of Survey Data** *by K. Jill Kiecolt & Laura E. Nathan*
52 **Using Microcomputers in Research** *by Thomas W. Madron, C. Neal Tate & Robert G. Brookshire*
51 **Stochastic Parameter Regression Models** *by Paul Newbold & Theodore Bos*
50 **Multiple Regression in Practice** *by William D. Berry & Stanley Feldman*
49 **Basic Content Analysis, Second Ed.** *by Robert Philip Weber*
48 **Models for Innovation Diffusion** *by Vijay Mahajan & Robert A. Peterson*
47 **Canonical Correlation Analysis** *by Bruce Thompson*
46 **Event History Analysis** *by Paul D. Allison*
45 **Linear Probability, Logit, and Probit Models** *by John H. Aldrich & Forrest Nelson*
44 **Cluster Analysis** *by Mark S. Aldenderfer & Roger K. Blashfield*
43 **Bayesian Statistical Inference** *by Gudmund R. Iversen*
42 **Using Published Data** *by Herbert Jacob*
41 **Game Theory** *by Frank C. Zagare*
40 **Microcomputer Methods for Social Scientists, Second Ed.** *by Philip A. Schrodt*
39 **Introduction to Applied Demography** *by Norfleet W. Rives Jr. & William J. Serow*
38 **Matrix Algebra** *by Krishnan Namboodiri*
37 **Nonrecursive Causal Models** *by William D. Berry*
36 **Achievement Testing** *by Isaac I. Bejar*
35 **Introduction to Survey Sampling** *by Graham Kalton*
34 **Covariance Structure Models** *by J. Scott Long*
33 **Confirmatory Factor Analysis** *by J. Scott Long*
32 **Measures of Association** *by Albert M. Liebetrau*
31 **Mobility Tables** *by Michael Hout*
30 **Test Item Bias** *by Steven J. Osterlind*
29 **Interpreting and Using Regression** *by Christopher H. Achen*
28 **Network Analysis** *by David Knoke & James H. Kuklinski*
27 **Dynamic Modeling** *by R. Robert Huckfeldt, C.W. Kohfeld & Thomas W. Likens*
26 **Multiattribute Evaluation** *by Ward Edwards & J. Robert Newman*
25 **Magnitude Scaling** *by Milton Lodge*
24 **Unidimensional Scaling** *by John P. McIver & Edward G. Carmines*
23 **Research Designs** *by Paul E. Spector*
22 **Applied Regression** *by Michael S. Lewis-Beck*
21 **Interrupted Time Series Analysis** *by David McDowell, Richard McCleary, Errol E. Meidinger & Richard A. Hay Jr.*
20 **Log-Linear Models** *by David Knoke & Peter J. Burke*
19 **Discriminant Analysis** *by William R. Klecka*
18 **Analyzing Panel Data** *by Gregory B. Markus*
17 **Reliability and Validity Assessment** *by Edward G. Carmines & Richard A. Zeller*
16 **Exploratory Data Analysis** *by Frederick Hartwig with Brian E. Dearing*
15 **Multiple Indicators** *by John L. Sullivan & Stanley Feldman*
14 **Factor Analysis** *by Jae-On Kim & Charles W. Mueller*
13 **Introduction to Factor Analysis** *by Jae-On Kim & Charles W. Mueller*
12 **Analysis of Covariance** *by Albert R. Wildt & Olli T. Ahtola*
11 **Multidimensional Scaling** *by Joseph B. Kruskal & Myron Wish*
10 **Ecological Inference** *by Laura Irwin Langbein & Allan J. Lichtman*
9 **Time Series Analysis, Second Ed.** *by Charles W. Ostrom Jr.*
8 **Analysis of Ordinal Data** *by David K. Hildebrand, James D. Laing & Howard Rosenthal*
7 **Analysis of Nominal Data, Second Ed.** *by H. T. Reynolds*
6 **Canonical Analysis and Factor**
C **omparison** *by Mark S. Levine*
5 **Cohort Analysis** *by Norval D. Glenn*
4 **Tests of Significance** *by Ramon E. Henkel*
3 **Causal Modeling, Second Ed.** *by Herbert B. Asher*
2 **Operations Research Methods** *by Stuart S. Nagel with Marian Neef*
1 **Analysis of Variance, Second Ed.** *by Gudmund R. Iversen & Helmut Norpoth*

To order, please use order form on the next page.

Quantitative Applications in the Social Sciences

A SAGE UNIVERSITY PAPER SERIES

$13.95 each

Place
Stamp
here

SAGE PUBLICATIONS, INC.
P.O. BOX 5084
THOUSAND OAKS, CALIFORNIA 91359-9924